World Rural Women's Day
15 October

GW00568962

ECO-AGRICULTURE: FOOD FIRST FARMING

ECO-AGRICULTURE: FOOD FIRST FARMING

Theory and Practice

Marthe Kiley-Worthington

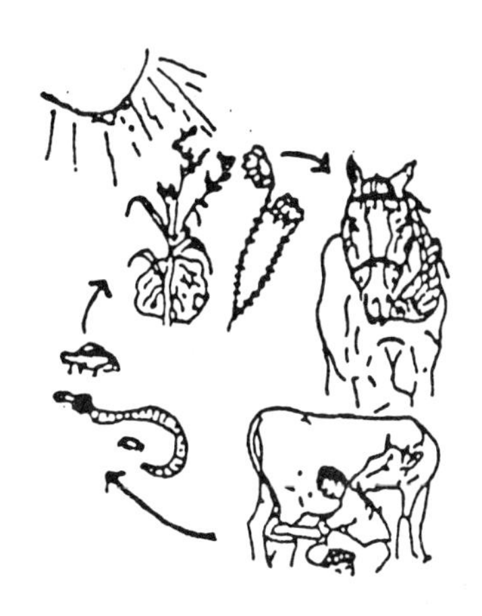

SOUVENIR PRESS

First published 1993 by Souvenir Press Ltd,
43 Great Russell Street, London WC1B 3PA
and simultaneously in Canada

ISBN 0 285 63117 9

Photoset by Rowland Phototypesetting Ltd,
Bury St Edmunds, Suffolk

Printed in Great Britain by
Redwood Books, Trowbridge, Wiltshire

CONTENTS

List of Photographs, Figures and Tables

PHOTOGRAPHS

FIGURES

TABLES

The case for a land ethic would appear hopeless but for the minority which is in obvious revolt against these 'modern' trends.

The 'keylog' which must be moved to release the evolutionary process for an ethic is simply this: quit thinking about decent land-use as solely an economic problem. Examine each question in terms of what is ethically and esthetically right, as well as what is economically expedient.

A Sand County Almanac: And Sketches Here and There. Aldo Leopold, 1949

To Jonas who taught me how to think about practical problems, to Rebecca who taught me patience (or tried to!), to Ekhukhanyani Secondary School, Swaziland, and all the other such bush schools in Africa, and to all those, now in the hundreds, who have given of their time, enthusiasm, sweat, energy and humour on the projects over the last twenty years – through good times and bad – and in particular to Ruth F., Tony C., the two Chloes little and big, Eric S., Charlotte D., Cathy S., Mike N., Nell Z. and Julian E., Vivienne L., Kate R. and many of our four-footed friends, including Fuzzy, Barney, Baksheesh, Sheeba, Aderin, Omeya and Crystal and their clans. To Pamela, Ballo and all the others who helped finance these projects.

Special thanks to Charles Mosse for his perception, kindness and loan of a room to write in, and to Rosie Brindley for help with the typing.

I am grateful to Methuen & Co. for permission to quote the dedication from *The Hungry Future* by René Dumont and Bernard Rosier.

Above all this book is dedicated to Chris, my partner and like-thinking pal, without whom most of this could never have happened.

M.K.-W.

Preface

Even as a small girl in Africa, I could not help being aware of what was beginning to happen there. Instead of food crops planted in clearings through the wattle forests, the trees were all being chopped down, to be replaced by small plantations of pyrethrum flowers and even roses. Children were numerous, but they and their parents rarely looked badly nourished. They were poor in the sense of having only what they could find or make locally, but took great pride in their tribal dress and ornaments.

As I grew up, clapped-out Ford trucks began to disfigure the village commons, cast-off European clothes became the norm and gradually, malnutrition began to show in the children as their fathers devoted more and more land to the cultivation of the pyrethrum and rose cash crops instead of the food crops that the women had always grown before. We lived between the Kikuyu 'reserve', as it was called, and the forestry, veterinary and agricultural research stations. I looked with wonder at the humpless cows that had been imported from Britain to 'improve' the indigenous stock of small, tough, humped Borans, but which had to be kept fenced and frequently dipped in order to survive. The experiment may have amused the veterinarians, but what possible relevance could it have for people like Mturi and his family?

Later, when I returned to East Africa to work on wildlife research, these changes were even more obvious. Certainly there was more money around, but there was also more malnutrition and less to eat; a few people had become very rich, but many more were starving in bad years, as well as having their lives disrupted by the political upheavals that have beset peasants everywhere throughout history.

There was something wrong with what was happening in agriculture, and perhaps also with the value systems that were being indoctrinated into those rural people. Why should they have to be dragged into the consumer economy? Why should they be taught to want more of everything, so long as they were comfortable, well-fed and happy? Why should they have to become dependent for survival on others? Perhaps this is why peasants have always had such a bad time: they are by nature independent, and those in power find this difficult to cope with. These African peasants could well have had something to teach the European consumers who had long since stamped out their own peasantry and replaced them with an urban-induced, industrialised agriculture. They had no need of credit, electricity, running water, motor cars or expensive houses, and they had a lot of laughter in their lives. Might there be other approaches to 'development' (whatever this meant), and if so, what, why and how?

Such questions began to exercise me even at the prim girls' boarding school where I always seemed to be out of step, always asking why I had to put on a hat, stand to attention, bow and scrape to older girls . . . because it was the Rule! I was indeed aware of that: but *why* was it the rule? What possible benefit could it have, and to whom?

At that time I had no overall plan, only an inability to live for long periods in an urban environment. Gradually opportunities opened to allow me to return to the questions that had first troubled me in Africa, and meanwhile the problems there, and in many other parts of the world, had grown exponentially. At last, about twenty years ago, I had a chance to become at least a partial rural peasant. Might this give me the opportunity to see if I could help solve some of the world's growing problems?

Introduction

A borrowed tractor hiccupped itself and a trailer-load of old furniture, wire-netting and children's clothes down the A27 towards our new acquisition: 17 acres of flooding pasture on the bank of the river Cuckmere, a 14-bedroomed, somewhat ramshackle baronial manor house and, most important of all, an original Victorian farm-yard, with coach-house, working horse stables and loft, granary, cattle loafing yards, separate flint-built duck, hen and geese houses, an outside privy in the orchard, and a picturebook flint-built pig shed and yard . . . fine for the human expectations, but suspect for the pigs. In addition there were several mature walnut trees, a chapel long since converted to a junk house, and a very ancient, very large, very tumbled-down dovecot from the days when baronial masters allowed hundreds of pigeons to nest, eat everyone else's corn, and supply their groaning tables with yet more fresh protein.

It was a cold November day, threatening snow. Syringa, the equine Cleopatra of my life and the original reason for this move, and Collywobbles, her companion and my sons' friend, were being led the 15 miles by our South African baby minders, Chaos and Matilda. Nothing like a good long walk to calm them all down and warm them up, I had thought. In any case, Anglo-Arab fillies, white South African young women and chaotic young men were not likely to be great assets in a farm-moving enterprise—better out of the way but at least feeling they were helping, which of course they were.

Sod's law, or bad maintenance, dictated that the tractor should break down at frequent intervals, and consequently we arrived after dark. The electricity was off as well as suspect, but a large, dark, perishingly cold baronial mansion was an ideal place for a game of spooks and frights, at least for those who were not consumed with worry and panic. There were fireplaces in the 15-foot-high Victorian rooms, and with a little more energy and ingenuity we gathered some fuel to make a fire—only to find smoke billowing into our shivering faces as we huddled close to pick up the slightest hint of warmth. It did not take long to identify the problem: not a chimney in need of sweeping, but a cemented-up chimney. Our predecessors had economised on electric heating by sealing up the chimneys!

Such was our introduction to Milton Court Farm in 1970. As I lay in my sleeping bag in the smoky, draughty cavern of a room, I began to reconsider my enthusiasm and my determination to avoid becoming part of the consumer economy, earning money to pay money. Why have all the worry? I had thought. Why not grow and supply the needs of one's whole family, animal as well as human? Or at least all one needed for a well-fed, healthy, relaxed and enjoyable life with time to think, read, stare and learn more about other living things. Such a life, which need not cause environmental problems either locally or globally, would also, it had seemed to me, be more ethically defensible.

As I lay doubting my decision that November night, shivers began to creep through my chilled body. The very obstacles kindled the energy and determination to muddle through somehow. Surely, if the Kikuyu women of my childhood could manage, so could I.

It is a paradox of Western society that the same people who deplore the destruction of the rain forests and the threat to the environment may also buy food that is the end-product of intensive farming, with slurry lagoons, nitrogen applications, a veritable Pandora's box of herbicides and pesticides, and animal production relying on food often grown where there was once rain forest.

The curious idea that modern high-input agriculture is 'efficient' and 'economic' has been successfully sold for far too long. Now countries like Ethiopia and the Sudan are reaping the results. Such human and environmental disasters could have been avoided if reason and thought had been the name of the game. In the 'developed world' we have tended to take the attitude that the problems are for other people, but even in 1970 it was blatantly obvious to me, and to some others at least, that the party was over. Somehow we had to put our own lives and environment in order before we set about 'developing' others and telling them how to do it all. My contribution would be to think about and practise a food-producing and animal husbandry system which faced the problems and tried to find some solutions. This was why we had made that cold, bumpy tractor journey to this downland valley.

The next morning, accompanied by Sneefus (a black four-legged canine friend with an onomato-olfactory name), I visited Syringa in her field. As the sun rose I looked across the Cuckmere valley to the gentle, rounding downs. There would be backaches, perhaps struggles, disasters, but I was convinced I had made the right decision; on balance there would be more joy than distress. There had to be—after all we were privileged to be alive! Looking at Syringa and Sneefus, their enjoyment of the moment coupled with my childhood culture of African optimism was a quick fillip to my enthusiasm in the face of discomfort and plain panic.

So what were the problems that confronted (and still confront) modern agriculture? What had gone wrong? Even at that time there were grumblings and murmurings, headed by the 'Muck and Magic Brigade' as Lady Eve Balfour's Soil Association was known. This small group included many thinking people from many disciplines. Were they just cranks (aptly defined by E. F. Schumacher as a 'small implement that causes revolutions'), the famous 'British eccentric', or were there serious messages produced by such people that must be taken on board by the establishment?

Much of this is history now as the general public have forced some change and some rethinking. But there remains a powerful majority in the agricultural establishment, which seems incapable of admitting that it was wrong and of considering the need seriously to restructure agricultural education. Sadly, this group includes not only those whose wealth is dependent on supplying the high-input, government-sponsored agriculture, but also many intellectuals in research and university teaching, who have neither heard nor understood, never mind seriously considered, the messages concerning modern agriculture's appalling muddle and how we might get out of it for the long-term future.

Back in 1970, my first task was to identify by serious research where the problems actually lay. Was agriculture really 'efficient'? Was it really economic? Had it progressed and used biochemical, physiological and above all ecological knowledge appropriately? What was appropriate anyway? Was and should agriculture be an 'industry'? What was the point of it all?

The more I researched the theory, the more depressed I became.

There was, and still remains, a galaxy of problems confronting agriculture. However, it is always much easier to be critical and derogatory than constructive. My aim was to try to overcome these problems by finding a workable alternative. This sparked off the germs and then the evolution of Ecological Agriculture, what it is and how it might work.

My parents' nearby farm illustrated only too well some of the conflicting land interests of the late 1960s. My mother, the farmer, was deeply concerned with paying the bills and with economic survival. Also, perhaps, in line with the beliefs of the time, she measured her self-respect against her economic success in modern dairy farming. She was a woman in a predominantly male world, and was determined to succeed by men's standards. She built up and ran an efficient, high-yielding, high-input dairy herd of pedigree Guernsey cattle, and became one of the most knowledgeable and successful grassland improvers and growers in Sussex.

My father was the first scientific director of the Nature Conservancy Council, an ecologist, one of that early breed who altered the world view, established the International Union for Nature Conservation and helped to set up nature reserves in many countries. He was a dedicated conservationist, but, as with other ecologists of the time, he saw the world of the agriculturalists as alien but necessary; he assumed that they knew their job and could do it better than anyone else. So although he left my mother to get on with the farming, he made demands on the land too. He enjoyed shooting and fishing and wanted woodlands and lakes for his sports, and cover in hedges for his pheasants; if his sports were to be affected, then chemical additions to the land must be ruled out.

Both of them could have had their land needs fulfilled by developing an Ecological Agriculture. But they did not, and some woodland had to go for conversion to grassland so that there could be more cows to pay the bills. This was a valuable chestnut coppice, but in line with current opinion both my parents believed that they could not manage it economically. Reducing woodland also reduced the number of pheasants that could be supported. Then some land was taken from the farm to be made into lakes for trout. The farm suffered its loss. Then the silage effluent from the farm polluted the lakes and killed the fish. So it went on. Surely, I thought, there must be a better way of food production which could integrate all these land use requirements, and more.

How was I to run my mini-farm? The prime concern was that it should be FUN. I had had too many 4 a.m. risings to milk 50 cows in the dark and frozen cow shed, followed by the desperate rush to heave the churns to the end of the drive before the milk lorry arrived, wash the machinery, bucket-feed the calves, move the electric fence, wash down the cow shed, make up the records, feed the heifers, start the tractor to scrape the yard . . . before starting milking again at 4 p.m. One finished at 7–8 p.m. if one was lucky, having worked against the clock, one's mind constantly occupied with trying to work out how to increase efficiency, so that bringing in the cows would take only four minutes instead of eight. One dropped exhausted and half-asleep into a chair before hobbling off to bed in order to get up at 4 a.m. the next morning . . . I had had enough of this to last a lifetime. Did farming always have to be a time-and-motion study, and an exercise in endurance and energy?

But my farm of the future must make some money, or at least cover its costs. It must also provide facilities that we wanted—for example, a living for my horse, and an environment where my sons could adventure and suffer the slings and arrows of growing up without the dangers of being killed by unnatural disasters such as chemical poisoning. Natural hazards like drowning in rivers and ponds, falling out of trees and competition between brothers were enough, and part of growing up since we all began. Dying of poisoning from the various agricultural chemicals that lie about in large quantities on every high-input farm were not. In the nature of things we were unlikely to be more careful than any other farmer about putting these away in locked sheds.

The farm must be a beautiful place; it must provide both our spiritual and our physical food, and as many as possible of any other requirements that we might have, so that we would not have to hurry off to make money to buy food and other necessities elsewhere. Then again, it seemed foolish that one farmer should have many acres or hundreds of cows which he struggled to maintain with large machinery and a burden of worry. The countryside and even gardens showed the results of this development—hedges rarely cut and laid, beautiful old walled gardens that used to provide food for the large households either overgrown and forgotten, or converted to neat lawns and flowerbeds, while the owners rushed out to buy aubergines, green peppers and pawpaws from the local supermarket. Old woodlands that used to be coppiced were being donated to the bird protection societies who were poisoning the chestnut coppice roots, while the farmer bought in fence posts grown and fashioned from the dwindling wild forests—in Australia and California.

At the same time there was a large section of the population infected with the euphoria of the 'Flower Power' era and a re-examination of the materialistic values they had inherited, who were searching for life-styles that would allow them a greater level of self-determination and scope to ponder on the meaning of it all. There was a veritable mob of skill-less, moneyless, well-meaning, sometimes energetic and often young people who were anxious to have at least small parcels of land to try out their ideas and live on their own resources.

The more one thought about it the more absurd current agriculture appeared to be. There were the commercial farmers sweating and worrying desperately, having their system maintained by tax-payers so that a few very rich people might grow richer, cutting every corner to save on expensive labour and so threatening the long-term future of the land and all its greatness and beauty. At the same time there was a large number of people with time and a desperate need and energy to learn how to live harmoniously outside the cities, unable to find a place to do it, and not even allowed to help the overworked industrialised farmer. Was there a way in which these could be integrated? Would it be possible to run a biologically and economically efficient small farm which would have a higher labour demand yet would avoid the grind and slog of the nineteenth-century peasant because, as one of my students put it, 'our heads are in a different place'?

Thoughts like these must somehow be moulded together into a food first agriculture. It would be nice to say that the whole thing was thought out in advance, that the exact numbers of stock, the cultivation procedures, the incomings and outgoings were designed and adhered to in a nice, neat, empirically motivated way. It was not like that: the farm evolved and changed as one's ideas evolved and changed, as people came and went; and as we went, we learnt.

Over the first ten years there were human births, marriages and divorces, grand passions and at least minor crime, great joy and great gloom, great good luck and periods of frightful misfortune;

moments of great triumph and terrible accidents; mothers grabbing daughters from their lovers' arms in the middle of the night, and mothers dumping daughters as a last ditch. I suppose we had most of the human story; but just as the humans went about their living and dying, so too did each of the animals whose lives were little less eventful. We were privileged to share each other's lives and gradually, I hope, evolved a method of living that was of mutual benefit to all. We made mistakes, but we learnt.

The next six years began with a chaotic move from Sussex to Mull in the Hebrides, to try out our system in a marginal area: to see if it would work for less privileged regions. Again, we had dramas, melodramas, great good fortune, great disaster. It was very hard work in a difficult place, but energy, good humour and positive thinking kept us going. Much of this was supplied by the beauty of the natural environment: the skies of unbelievable colour changes, the constant dynamism and melodrama of the weather, the endless sweeps of temperature winter and summer. Four years ago we began again in Devon, with just a piece of land to develop an ecological farm in a more purist form.

This book summarises the thinking, investigating, discussing, arguing and practical results of our ideas during those 20 years.

1 Sudan, Bread Basket or Bread Sink?

Over the last few years we have heard a great deal about the starvation of the people of the Sudan and Ethiopia and, more recently, Mozambique and Somalia. The pictures on our television screens, of babies with distended stomachs and fly-ridden faces, have pricked the conscience of the rich and overfed West. Bob Geldof managed to raise more for this cause than has ever been collected before.

The majority of us think that by giving generously the problem will be solved. All will be fed from the wheat mountain in the West, and we can go back to our self-obsessed easy living. Alas, it is not quite so easy. What is all the money to be spent on? Is feeding the growing population in these countries from outside resources going to solve the problem once and for all? What effect will it have on the distribution of resources throughout the world, and what will be the long-term effect on all of us?

To start with, we must all try to understand why the problem has arisen, then build up rational arguments and devise practical alternatives to decide what can be done about it.

Actually, I don't think we need to construct complex 'scenarios' of the different solutions. For many reasons, I believe that there is probably only *one* solution which in the end will help us all. This solution is simple: it is embodied in the adage, '*help thyself, but not at the cost of others or the environment.*' This applies as much to those in the so-called 'developed' countries as to those in the so-called 'developing' world. If we are to provide a constructive, sustainable future for humans and the biosphere they depend on, we have to recognise that, whether we like it or not, biologically speaking we live in a global village. How we live, what we think and what we possess may vary in small details between households, countries and species, but the basic controls of life and its infinite sustenance are the same for all of us—cabbage, maggot, lion or human; African, Chinese or Anglo-Saxon. For the future, we must recognise and build our diversified life-styles around these rules.

Much as the economist would like us to believe that we are controlled and must live by his rules, the sociologist by hers, the politician by his or the religious fanatic by his, the only rules that fundamentally control our survival or extinction, even our happiness, are biological ones—being alive or dead, sick or healthy, starving or well fed. The voting public should be aware by now that those presently holding maximum power in human societies, the politicians and economists, will not be able to continue to supply. Left and Right, they still promise us more of everything and an ever-rising standard of living, believing that this is what we all want and that it is possible: another economic boom, more and bigger cars, an even greater variety of foods, the ability to travel faster, farther and more frequently, more high-tech gadgets, and so on. The continual emphasis on consumerism, coupled with increasing human populations, cannot indefinitely be sustained; there are limited resources and major environmental controls that people in general, particularly in the 'developed world', have not seriously come to grips with.

The party is over, folks. We cannot continue to pretend that 'further economic growth, more money, expansion, increased industrialisation' and so on are either possible or desirable. Why should those of us who, in the modern 'developed' state, lack nothing to sustain and enjoy life automatically assume that the only way to happiness is *more* luxury, more goods? Judging by the growing number of people in these countries suffering psychological illnesses, it evidently is not. It may be that we need to think far more seriously about our psychological health rather than concentrating on material gain . . . the two are not necessarily related.

In 1983, before the Ethiopian and Sudanese famine had hit the headlines, I was asked to categorise and edit a bibliography of all the research on agriculture and agricultural development in the Sudan, Jordan and Syria, from 1974 to 1983. The result, needless to say, was that I became well acquainted with all the reported agricultural research that had been done, and the fundamental strategy adopted and implemented by the West on agricultural development in these countries.

As I read the papers and pored over the massive, mind-boggling costs of the grandiose schemes, I became more and more critical. Why did the agricultural elite not pause to question what they were doing and how they were doing it? A charitable explanation was that they had not read the literature! A less charitable one was that they were primarily interested in feathering their own nests. Nearest to the truth, perhaps, was that the FAO (Food and Agricultural Organisation) and other decision-makers and their advisers had had the disadvantage of a specialist education in the prevailing dogma: 'being richer will make you happier' and 'we can achieve this by marrying economics with high-input agriculture'. Holistic, ecological approaches and good multi-disciplinary science were discouraged.

Unsolicited, I wrote a critical but constructive editorial for the *Agricultural Bibliography of the Sudan* (Zahlan, 1984). It was perhaps too controversial at the time and so was not used, but it helped me to clarify my thinking and produce examples.

Ten years on, it is fashionable to mouth Appropriate Technology, Ecologically Sound or Sustainable Agricultural ideas for Third World countries. But *what* does this mean? What changes should take place, and how? Exactly what has gone wrong?

Agricultural Mistakes

It is quite clear that there is a growing need for a multi-disciplinary approach to agriculture and development. Agriculturalists must be aware of the social, medical, veterinary and ecological effects of their work if, in the end, the indigenous population is to benefit.

During the period 1974–83 a staggering amount of money was spent on agricultural research in the Sudan. For example, there were 29 major FAO projects started in this decade – some 1,800 research reports emphasise this.

The projects and their development have been unsuccessful and have not fulfilled their aims. In 1975, for instance, it was forecast that the Sudan would be the 'Bread Basket for the Middle East by 1985' (Kilner, 1975). Instead it has become the 'Bread Sink' by being held in the ghastly grip of massive famine. In such areas it is becoming increasingly obvious that it is essential to have a good grasp of ecological principles, and to take into account the delicate ecological balance of the natural living system before causing gross changes to the habitat, however attractive the economic returns might look on paper. In addition, the sociology of the indigenous people must be understood in depth and catered for if large-scale agricultural projects are to succeed.

In the Sudan the availability of capital from international organisations and banks (FAO and the World Bank) led to the development of grandiose and very capital-expensive irrigation schemes. The failure of these projects to reach their production targets must lead to a serious examination of their relevance in such a country. For example, despite enormous capital input ($400 million), the Kenana sugar growing scheme, started in 1973, in which 170,000 hectares of arid land were to be irrigated to produce 300,000 tons of white sugar per annum to sell to other countries, only produces sugar for the home market (Wilson and Doran, 1980). The Rahed scheme, in which large areas where irrigated to grow groundnuts and cotton, fell apart because there were no spare parts, no labour and rather low yields—the wrong crop in the wrong place (Heinritz, 1982).

Even the Gezira scheme, a cotton growing project started by the colonial British in the 1930s, on which many other schemes were modelled, has been considered a failure by some (Barnet, 1977; Pollard, 1981).

Then the Jonglei project, to drain the Sudd (the vast swamps of the Nile in the south of Sudan) and irrigate other areas, has also been a failure. The Nubians who were given land there after being displaced by the formation of Lake Nasser following the building of the Aswan dam, did not want to farm there, and the nomadic pastoralists who had previously lived off the land refused to get rid of their livestock.

Despite encouraging statements concerning the possibilities of the Jonglei project (Mohamed, 1978) it would appear that the environmental and social costs were too high (Boyles, 1980; Critchfield, 1978).

A German publication published in 1979 states blandly the belief that was, and unfortunately still is widely held by 'developers' and those who research and advise: 'To achieve an acceptable rate of economic development, and to sustain that development, the Sudan must expand food exports despite the existence of unsatisfactory nutritional conditions among its own people' (Oesterdiekhoff, 1979). This assumes that the population would prefer to be rich, even if they are dead . . . a rash belief. A more thoughtful approach and less emphasis on economics might have steered away, or at least reduced, the famine in the Sudan during the last decade and a half.

It has been considered crucial to settle the pastoral nomads in the Sudan . . . no reasons given (e.g. Heinritz, 1982). Is it not time to ask *why* they should have to give up their livestock, or settle at all? In this arid, delicate, ecologically balanced tropical biotype—the dry savannah of much of the Sudan—it may be that the nomads achieve greater net sustainable production, and that pastoral nomadism is the most appropriate use of the land. However, growing populations inevitably put a strain on this. Research effort might well be spent on detailed studies of the nomads, the ecology and behaviour of their animals, their movement patterns and the carrying capacity of the land, before dismissing this form of land use, or suggesting that the problems are those of 'lack of capital and of the transport system' (Salih, 1978).

Another major problem which really is not properly understood and does not seem to be receiving enough research effort in the Sudan, or in many other tropical or sub-tropical parts of the world (see discussion on Australia in chapter 11), is increasing soil salinity as a result of prolonged irrigation.

Meaningless statements like, 'it is underdevelopment' which prevents progress and therefore the Sudan's aim to be the Bread Basket of the Middle East, are still being made (e.g. Kiss, 1977), but there is also a hint that things may at last be changing—although at the cost of many lives. Several authors have been muttering that a multi-disciplinary approach, or a food first diversified approach to agriculture, might be more appropriate. Some even suggest that traditional and 'modern' agriculture should and could be complementary rather than competitive. Even in 1978 it was being mooted that the idea of imposing western technology on developing societies could be inappropriate.

Sudanese agriculture appears to be an excellent example to illustrate this point: the time for imposition of western technology and the inevitable accompanying control by the West (or other countries playing a similar role) is over. To talk of raising the people from 'poverty and backwardness', as has so often been stated, is no longer appropriate—if it ever was. What is needed now is the development of self-reliance, particularly in food production, and hence not only the *appropriate* technologies to do this, but the appropriate *mind set*: food first, not money first. In addition, the development towards a mutually beneficial 'symbiotic' relationship between cultures, rather than the domination of one and the elimination of the other, might well be the best way forward.

It is evident that there has been no standard overall plan for agricultural development in the Sudan (Tewfik, 1977). This may prove to be a blessing in disguise when the record in agricultural achievements over the last two decades in the Sudan and elsewhere in Africa is considered.

Righting the Wrongs of Agricultural Development?

What different approach can there be, and what technologies are needed for it? Small projects to try out different systems in different areas might well be more successful than grandiose schemes. One of these was the Abyei project, an 'integrated rural development project' (Huntington, 1980), involving the use of alternative technology and low capital input, while retaining the indigenous social structure and patterns of behaviour. This also had problems connected with international bureaucratic organisations. It may be that Nyerere's 'self-help' ideas are really the solution. Peasant populations throughout the world may have some of the answers—greater self-reliance, thus political independence and a great deal more practical skills than the 'development' advisers or researchers ever had.

A self-sustaining low-input, high net-yielding agricultural system, requiring low capital investment and with an emphasis *firstly on food crops* and secondly on cash crops—*Ecological Agriculture*—might be more appropriate. Much of the technology for such a system is available, but certainly it requires further study and demonstration farms to be set up in such countries.

There is still much research to be done on the physical environment of the Sudan, particularly on climate and vegetation. The development of self-sustaining agricultural systems will depend on the most detailed information since, by their nature, they change to fit the existing natural conditions. The Sudan is a large country whose basic ecology appears to have been little studied, even though it is from a comprehension of this that successful land use and food production planning will emerge.

How is Ecological Agriculture Practised?

The development of a food first agricultural strategy rests heavily on being able to increase the net biological production from agricultural land. The net production is calculated by subtracting from the gross yield the inputs applied from off the farm (p. 61). Dependence on nutrients from outside the system or country requires money and often contributes, firstly, to the food shortage within the country or area (see p. 71), and secondly, to environmental problems in either country. The first principle, then, to help towards obtaining a sustainable food first agriculture is to find and apply organic wastes from any source, including industrial by-products, to help increase the decomposers and thus improve the soils. The farm-yard manure of the Sudan and Sahel is often burnt as fuel for cooking, rather than being used in this way. This is because in areas where the majority of the human population lives, there is no other fuel since deforestation is almost complete. The more the trees are cut and the manure burnt, the worse the situation becomes. This is what is called a 'down-grading ecological system'. If it is to reverse to an 'ecological upgrading system' (see p. 61), then organic wastes must be recycled, something that has not been seriously considered in the Sudan, or in most countries. For example, the spread of the water hyacinth (*Eicchhornia crassipes*) could be considered a bonus and be harvested to act as a mulch on the dry soils with low humus, yet to date it has been regarded as a pest and treated with herbicides, resulting in further environmental problems: polluted water and no lasting control of *Eicchhornia*.

The Forests

Reafforestation, considered by many to be one of the key factors in beginning to ecologically upgrade an area, is essential in much of the Sudan, as in so many other countries. Schumacher (1974) calculated that if every member of the human population in India were to plant and look after a tree, the food, fuel, shade and wood shortages would be solved within ten years. His advice was not heeded, but it would still be possible in many countries . . . even the Sudan.

The west of the Sudan is almost unique in having large areas of indigenous hardwood forests with a great wealth of species. It must surely be one of the major resources of the country. However, to avoid the evils of deforestation and resulting land denudation that have occurred only too frequently in other countries, it is essential that the forests are properly understood and managed. Yet the western-financed forestry research has, as usual, concentrated on encouraging the planting of introduced species, such as quick-growing eucalyptus from Australia, rather than the recognition, let alone the management, of the indigenous species. This does not bode well for the future of this invaluable natural resource, despite the fact that the Sudan is one of the major world producers of gum arabic, used in the manufacture of glues and in the food and pharmaceutical industries, which is produced from an indigenous hardwood tree.

Quick-growing, fire-resistant exotic species such as *Eucalyptus spp* may well make an important contribution to the economy and might also help with soil conservation, by acting as shelter breaks or being cut and layed as hedges, but the prime concern must be with the management, conservation *and* sustainable use of the indigenous species which have evolved to live and reproduce in that environment, with the development of limited markets for the natural forest species accompanied by a proper replanting programme.

Integrating trees into the farm as shelter belts, stockproof hedges, for cropping as coppice trees, for fruit and fodder throughout the farm rather than only in the forest or farm woodland, allows for multi-purpose land use and aesthetic and environmental improvement for the people who live there (see pp. 95–6). There is nothing like a few shady trees to sit under in the hot tropical sun, whether you are a human or a cow! There are many rural technologies in different parts of the world that make sustainable use of a wide variety of trees in many different ways. Much could be done to solve food problems if there were an exchange of this knowledge worldwide, accompanied by motivation of the populations in the original peasant and isolated community way of self-help, rather than the recently introduced dependence on others, which results in people waiting for hand-outs. Let us hope that during the next decades there will be a change in emphasis towards sympathetic research and management of indigenous forests and the planting and maintaining of many species of useful trees outside forests.

Desertification

Desertification as a result of deforestation is a major problem in Africa, Asia, Australia and America and is the result, generally, of efforts to increase gross production of, for example, cash crops, sheep, goats, sugar or cattle. It also results when there are just too many people trying to live and cook off one area.

Water and Irrigation

In a country where agricultural inputs are growing exponentially and where, over much of the land, water is the scarcest of vital resources, the run-off of fertilisers, pesticides and herbicides into waterways is likely to become a major problem. For example, it is possible that it is nitrogen run-off from fertiliser application on the land that has caused the rapid spread of the floating water hyacinth which blocks up the flow of rivers and cuts down oxygen in the water. This in turn results in the death of the indigenous river fauna and flora.

Water pollution can have spin-offs in unprecedented ways. For example, the spread of the parasitic worm which causes bilharzia, one of the most widespread major debilitating diseases, might be, at least in part, the result of water pollution killing mollusc-eating fish. The worm has two hosts, humans and a specific snail; with no predators to control their numbers, the snails increase and, with them, the bilharzia.

There is also concern over the possibility of contamination of waterways with herbicides. Nevertheless, applications of 2.4.D. (a herbicide that kills all species and does not degrade) are still recommended, despite the problems this chemical has caused elsewhere in the world.

If the water were cleaned up, fish production could be an important self-sustaining protein food. During colonial times, 40 years ago, the infrastructure (research on the ecology and population dynamics of the fish stocks, catching techniques and processing) for this was well established (Tothill, 1948; Worthington, 1946). There could be an immediate spin-off in providing an excellent source of cheap protein at little or no environmental cost.

With the development of irrigation goes the spread of bilharzia, onchocerisis (river blindness) and malaria. These diseases are essentially environmental and will only be overcome by detailed studies of the ecology and ethology (behaviour) of the parasites and their vectors. Such studies are long overdue. It is difficult to see that there have been any major developments in the cause, control or cure of these diseases during the last decade; however, to be fair, progress is inevitably slow in such areas. Meanwhile it might be constructive to consider the role of traditional folk medicine which is proving possible and useful in other parts of the world.

Malnutrition

The severity and size of the problem of malnutrition can be judged by Taha's finding that 47 per cent of the children in the Gezira (a relatively prosperous part of the Sudan) were badly nourished as long ago as 1978. It could be that where agriculture is food-orientated rather than cash crop-orientated, malnutrition is not so great. It is unwise to measure poor nutritional level by *per capita* income. The two are not necessarily related in food first agricultural systems: self-sustaining peasants or pastoralists can be well fed although they have no monetary income. This point has not been understood by writers to date (e.g. George, 1976), with the result that agricultural and nutritional improvements have been geared to increasing *per capita* income rather than net food production. The theory behind increasing the gross national product is that the people will not be badly nourished, since they will have money to buy food. The truth is that if there is no food to buy, they will still starve—richly!

I can't help feeling that if only there had been a few peasant food-producing women in advisory councils worldwide, the development of the agricultural industry as primarily a money-making concern would probably never have been so universally embraced, and agricultural developments would have gone along a very different route. But it is not too late.

Plants

The variables we need to know more about for self-sustaining plant husbandry are such things as the best planting time, spacing and design of planting, interplanting and companion planting, and rotations. Since there is a growing world fertiliser shortage and the environmental, social and monetary costs of chemical fertilisation are escalating, food and fodder legumes (nitrogen fixers) have vital roles to play in food first self-sustaining ecological agriculture.

Another field for development, but with an emphasis very

different from the one that has usually been given to it to date, is plant breeding to develop hardy, disease- and pest-resistant, higher yielding varieties which can withstand sub-optimal environmental conditions. Saline-tolerant species are of particular importance in the irrigated areas.

One indigenous plant, Sudan grass, is used worldwide as a fodder crop, but it is little used in the Sudan, despite its quick and massive growth. Surely it has potential in its country of origin, particularly for animal fodder, as hay or silage.

The Sudan's climate and its irrigated areas with alluvial soil are ideal for vegetable growing. One of the major incentives for the production of horticultural crops is the large amount of human food that can be produced from a small area (Griffin, 1974). Vegetable production by and for individual households would be a way of providing food, improving diets and, possibly, once these priorities are fulfilled, a way of earning foreign currency by export to other countries. Some of the vast areas of irrigated land used for sugar or cotton growing could, one suspects, be more efficiently (and possibly even more profitably) used for small or even larger-scale vegetable production.

With irrigation, Sudan produces citrus fruit for sale. Self-sustaining food first eco-farming would concentrate on fruits such as dates which are a high energy, very palatable food for humans and animals, able to grow with very low rainfall.

If herbicides and pesticides are used or recommended (there might be a case for either or both in certain areas), then to avoid environmental problems which are so far unidentified, it would be wise to use an integrated approach. This is where the natural local predators, natural weed controls and so on, become important components of the programme.

The use of indigenous plants by the indigenous populations has not as yet been seriously worked out; rather, exotic plants are introduced to replace them. This may often be misguided, since local plants are usually best adapted to local conditions, and the local people often have much knowledge of them. One exception is the proposed new idea for using papyrus (a fast-growing tropical sedge) to recycle wastes (Gaudet, 1979). Papyrus could be more widely used for the construction of houses, shelters and boats. It was traditionally used by many people throughout tropical Africa, living near lakes and rivers where it grew.

Animals

Large sections of the population are pastoralists, relying entirely on livestock. The numbers, distribution and performance of these animals are apparently unknown. The carrying capacity of different areas has been roughly assessed (Ali, 1976), but the feed preferences and total intake (how much each animal eats) of the different species are not known. Only by researching such facts will it be possible to calculate how many animals of each type can be sustained. This is vitally important where there are nomadic pastoralist animals and humans.

Unfortunately the majority of research on livestock has taken the West as a model and concentrated on feeding experiments in confined feed lots, often of isolated animals. The relevance of such work is hard to see. Intensive livestock production in an area such as the Sudan seems particularly inappropriate, if it is appropriate anywhere, which is doubtful (see chapters 8 and 9). Some of these experiments have even been done on imported *Bos taurus* (European cattle) instead of the more disease-resistant hardy local cattle, varieties of *Bos indicus*—cattle with humps. The completely inappropriate education of agriculturalist advisers and researchers in the West by teachers who rarely have a grasp of the realities (and certainly have never questioned whether their basic philosophy is appropriate), must be blamed for this.

1 *Dairy production and products*

There is little knowledge in many parts of Africa of how to preserve milk products by making cheeses, ghee and butter, for example. Often, therefore, the sour milk is thrown down the drain. A group of home economics students from 'developing countries', who were attending a one-year course in Britain, presumably learning how to run a house the 'modern' way, once came to visit our Sussex farm and see the dairy. Despite the fact that three of them came from dairy farms established where no one had had a dairy before, none of them had been taught what to do with milk—only that it was imperative to have a refrigerator and deep freeze. I wonder how everyone managed before these things were invented, or when there was no electricity. This has been forgotten in Europe. The emphasis is on how to provide the electricity and the freezer, not on how to preserve the milk. There are traditional methods of storing meat using locally available resources such as salting, sun- or shade-drying (biltong in South Africa, for example), which are often more appropriate than the expensive and technologically elaborate process of freezing, yet it seems that no one to date has thought about research in this field. Yet another area where food first agriculture should change the direction of research.

2 *The goat*

The goat, an amazingly opportunistic survivor in denuded arid zones, has had bad publicity as a creator of deserts. Despite being the prime candidate for research, it has received very little attention, largely because it was relatively unknown by western agriculturalists. No agriculture courses taught 'Goat Husbandry'. Since the goat has been the last mainstay of many human populations, particularly on the edge of deserts, I find this extraordinary. The first step when trying to help the agriculture in a country such as the Sudan would be to study the goat.

3 *The camel*

The ecology, management and production of the camel, another animal beautifully adapted to much of the region, is little known. It is a phenomenally useful and able animal, providing the energy for working the land, transport, fibre for clothes, milk and dairy products, and meat; again, largely ignored by agricultural research.

4 *The equids*

The equids (horses, donkeys, asses and mules) are extremely important for transport and, in the case of horses, for social status in this part of the world; yet again, very little is known or has been researched on how they fit into the system, what they eat and what other effects they have on the area. Donkeys have the advantage of surviving almost everywhere, and could become particularly useful draught animals to help the women with their cultivation. So far there are Intermediate Technology projects training donkeys and men to use them, yet it is women who do and will continue to do the digging! No one has thought about a woman training the donkeys and the women.

5 *Wildlife*
Wildlife is another possible source of food, fibre and other products which can be managed in a sustainable way and can integrate with human populations, but again in the Sudan there is no effort to try and integrate it with agriculture (see chapter 7, p. 151). It is recognised in much of Africa now that the wildlife can be a real and important source of food and raw materials and that local populations can utilise the environment often more efficiently than can introduced species, requiring less money to be spent (Dasmann *et al.*, 1973). Understanding the wildlife, its ecology population biology and usefulness is probably the best, and in the end the only way in which all the millions of species in Africa are likely to survive. The quicker this can be done and understood by the local populations, the more chance the elephant, the rhino and the countless numbers of other less well-known species have of survival. Integrating animal and human interests is the answer, not separating them as the misguided animal savers encourage (see p. 149).

Appropriate livestock are those that are adapted to local conditions and do not compete with humans for food. Examples of inappropriate animals are imported western high-producing breeds of, for example, cattle and poultry that will *only* survive with the aid of very careful husbandry and a battery of medication. Such animals also require high-quality food. In the Sudan, twenty times more research was conducted between 1974 and 1983 on imported cattle and poultry and how to manage them in intensive systems than on the local animals and plants. The approach continues throughout Africa still. If only one of those highly paid researchers had bothered to ask one of the local peasant farmers what questions he or she needed answering, millions of dollars would have been saved, and possibly millions of lives.

The misplaced euphoria concerning the high-yielding plants or animals must be resisted in favour of breeds more likely to survive although less spectacularly productive. These are normally bred from indigenous stock. For example, the backyard fowl in the Sudan has a valuable place as a converter of waste scraps to animal protein in the form of eggs and meat (see p. 212).

Sociology and Infrastructure

Labour problems, shortages and demands for higher wages have resulted in increased mechanisation in the Sudan, as in most other countries (see p. 47). Rurally based small-scale cottage industries using local, sustainably produced raw materials must surely have a major place in the future, now that the problems of large industrialisation are recognised.

The role of women in rural development in Africa is crucial, since they are usually the food producers and preparers (see chapter 10, p. 218). In the past it has been men and usually expatriates who have educated men in food production—it is not surprising, therefore, that change has been slow in Africa. This may indeed have been a blessing in disguise, since often the inappropriate 'modern' developments have not filtered down to the practitioners—the women. 'Development' has not led, either, to the increased economic independence of the women, but in some cases to a decrease (Benson and Duffield, 1979). It is to the women that suggestions and help in the development of food first agriculture are likely to appeal most and be of most use. The development of teaching and demonstration experimental ecological farms in the Sudan (and elsewhere in Africa) *for women by women* would help enormously in this direction.

Agricultural education by the use of television and radio broadcasting is widely used. This has enormous possibilities . . . provided the people are taught the appropriate things—but what are these? One particularly inappropriate common solution is to send the bright young men on grants from the Overseas Development Agency and other similar agencies to study agriculture at Western universities where they learn to develop further the cash crop economy and all sorts of quite irrelevant techniques. Then, on their return, they put it into practice, becoming advisers, teachers and researchers. I well remember meeting recently several Africans in the Department of Agriculture at the University of Edinburgh, who were doing just this. One from Ghana was working on very large high-input dairy units and their computerisation, and he himself felt bemused and lost. It was a pet project of his professorial supervisor who had other 'very important things' to do. Another, equally unsure of its relevance, was learning elaborate biochemical techniques in microbiology, while his country starved. He was concerned but dared not make a fuss. Another was learning all about intensive poultry and pig production, even though his country was Muslim—perhaps the most inappropriate information in every respect. I saw the results of this on one trip to Egypt where, with great pride, they showed me the new intensive poultry farm factory in the desert. A recent graduate from a British university had been given the money to establish it. We waded through the dead hens (the air conditioner had failed) and saw the piles of high-protein hen food imported from the United States to feed them. Each bag could have kept a few people there a little better fed than they were at the time.

Of course everyone can quote and excuse a few misguided projects. This is not my point. It is that the entire modern concept of agriculture in the West is misguided, particularly for developing countries. It is time Europe and Britain in particular gave up their BSc, MSc and PhD courses in agriculture for other countries—at least developing countries, where people's lives will be at stake unless they can do a very much better job than they have usually done. There is of course no reason why students should not come to study in Britain, but let them study other subjects.

There are other approaches. The Ujama concept (developed in Tanzania) of self-help and self-reliance may be relevant in many Sudanese communities, a meeting of new and traditional ideas and techniques for the self-sustainable use of local resources and appropriate technology. Agricultural peasant technology tends to be very local and to reflect local needs and environmental possibilities. An exchange of these while also exchanging cultural ideas may be where real help in agriculture for both the developed and developing countries lies. In the Western world we have as much to learn as to teach.

To date either large-scale mechanisation for large holdings or traditional hand tools have been the alternatives. Even here aid programmes can go very wrong. Recently I had a student who had gone out to Burundi to help a local group build an ox wagon and drive oxen. First, at great expense of aid money, a road was built to carry the ox cart, then the ox was bought and trained, and the cart made by the aid agency. This volunteer asked what would happen when she left. The local chief replied that they would have a big party and eat the ox . . . and why not?

Encouraging small-holders, tribesmen and peasants to use their own initiative and the material around them to lighten their work is important. There have not, for example, been any studies on the use of renewable, environmentally sound alternative and locally available energy sources, such as solar and water power, or on draught animals, particularly donkey power for women (see p. 225). These are obvious areas in need of research and application. In some cases there is information available (for example, the sun-powered water pump made from backyard junk that I saw operating in Australia; and a windmill made from box bits and an old car alternator in Devon).

The attention and money that the 'infrastructure for agricultural development' has received in the Sudan reflects the emphasis on growing cash crops for export. Roads, for example, have received more than their fair share, particularly the road from Juba to Kenya which has many political overtones. Existing transport features, such as the Nile, could perhaps be better used. Storage, transport, processing and packing have all received attention, but matters like the provision of fencing using locally available materials, or living barriers (such as hedges of appropriate plants) to control grazing and keep marauding animals off food crops, have not been considered.

During the colonial era some basic research on agriculture and endemic diseases, forestry and fisheries was done and published (e.g. Keen, 1946; Tothill, 1948; Worthington, 1946). It is a somewhat humbling experience to find that, with a few exceptions, there seems to have been little progress in agricultural knowledge and development in the Sudan, and few new ideas, since Tothill published his classic work in 1948.

Perhaps the time has come to think of alternatives, since the conventional agricultural approach has not been successful in either creating a 'Bread Basket' for the Middle East or in even feeding the Sudan's own people. One of these might be to consider the indigenous agricultural and pastoral systems of real worth, but with our new understanding of, for example, soil science, integrated pest control, ecology and behaviour of humans and grazing animals; to help without destroying or changing this. Sending periodic food packets to assuage our own consciences will not help in the long run. The population simply becomes dependent on them and continues to grow, thus becoming less and less self-sustaining.

The first principle of the indigenous peasant agriculture and land management of a region is survival of oneself and one's offspring without endangering others' survival. This is achieved by a self-sustaining, high net-yielding *food first* agriculture, the cornerstone of which is self-reliance rather than capitalisation.

2 The Problems of Modern Agriculture

Agriculture, like other human developments, reflects the philosophy of the time and culture. There are three aspects of this, which, if we are interested in the causes of the problem and the possible solutions, need at least to be pointed out:

1 *The anthropocentric view*
This has its origin in the Judaeo-Christian approach to the world and living things in it. Man is seen as being at the centre and as having certain unique attributes: he is made in the image of God, he has a soul which no other living creature is believed to have, and the rest of the world is there to serve his needs. Agriculturalists have been carried away on the wings of their own successes in such things as mechanisation and plant breeding, and by the development of the agricultural industry to serve man at all costs. They have not questioned cultural belief or stopped to think whether other living things might also be worthy of some consideration. They have ignored the long-term consequences of their particular developments and, most importantly since they are dealing with biological systems, they have overlooked some basic biological tenets that could be very useful to them. This philosophical approach is still only too frequently encountered.

2 *The technological fix*
Emerging from this anthropocentric philosophy is a deeper and more worrying belief to which we have all been exposed in the West, and of which agriculturalists are some of the chief exponents—that humans can out-smart nature or evolution. This belief has been fuelled by such technological successes as putting men on the moon, the elegance of nuclear power production, the control and wiping out of smallpox, the breeding of the 50-ton milk-producing cow and of high-yielding varieties of wheat. Many, particularly within the scientific establishment, believe that

it is only a matter of time before we can find a technological fix for everything. Even the sting in the tail—the nuclear waste problem, AIDS, the short life and disease problems of the 50-ton cow, and the social and biological problems of high-yielding varieties of rice, wheat and so on—does not seem to cause the slightest dent in this anthropocentric arrogance.

3 *The evolutionary approach*

This belief is characterised by the central idea of evolution, that an individual will tend to behave in a way that maximises his own chance of survival and that of his offspring. This means using the world to serve these ends and results in egocentric behaviour, while altruism is regarded as acceptable only insofar as it will further the same cause. An acceptance of this belief is at the root of the current consumeristic philosophy—getting rich is good—which has now developed in some countries to such a point that social status, power and successful competition based on the amount of money accrued are recognised as desirable and are the chief ambitions of many people. So normal is this attitude that those who state that they are not interested in being rich or earning more money are simply not believed . . . they must be lying or making enormous sacrifices!

What is not reasonable, if one embraces this evolutionary philosophy, is to behave in such a way as to jeopardise one's own survival and that of one's offspring and relatives. If we are doing this by using up non-renewable resources and polluting others, causing irreversible climatic changes, then we are not going to help your offspring or mine to survive. This must at least be stated if not seriously acted upon.

I do not intend to write a philosophical treatise here, only to point out that there are certain belief systems that we take for granted, which should be questioned if we are to try and find workable, rational solutions to the world's problems.

It is a combination of these philosophies that has given rise to the food crisis and many other problems worldwide. In the case of food production, the result is that one portion of the world lives off the back of another—for example, the development of cash crops grown in the Third World, the 'South', to sell to the rich world, the 'North'; while the North causes major environmental crises and changes as a consequence of its industrialisation, including agriculture—such as acid rain, global warming, nuclear wastes, agro-industry pollution and so on.

Curiously, in 1985 farmers in Europe thought to assuage the niggling of their consciences by sending 'gifts' of their over-produced grain to the starving, when the agricultural materials, such as phosphates, were in part produced by the starving Third World nations in the first place!

It might be asked, what is the point of quibbling about all this? Should we anyway be interested in people on the other side of the world? There is general agreement that agricultural strategy as it has developed in the rich nations has problems—fairly major ones. But when it comes to food production, unlike the motor car industry or any other industry, these problems are not purely economic, they are not just about our 'standard of living', about luxuries. They are about living or dying, about food or no food. We have a chance of living without motor cars, fertilisers, fuel and money. None of us anywhere has a chance of living without food.

It can be argued that all the people around the world who are starving, and the future generations who are going to starve, should never have been born. This view has its points, and perhaps one of the major concerns must be controlling human and other animals' population growth. On the other hand, one ecological measure of over-population is a shortage of resources to feed and provide other essentials (such as shelter) for the population. If a population can sustain itself indefinitely then there is no over-population. It is true to say that many poor countries cannot sustain their human and animal populations. We in the 'developed' countries should not pat ourselves on the back, however, since *we* do not sustain our populations from our own resources, and we have not controlled our population growth, even if we constantly remind ourselves that we have successfully reduced our growth rate to one per cent. This still means an increase of several million per year to feed, and a population doubling time of 60 years or so.

As I became more involved in finding out what agriculture was up to and what developments were going on throughout the world, a dusty cobweb of interconnected problems began to emerge. The next task was to try and understand their fundamental cause. The economist will see them one way, the animal producer another, the animal liberationist or the philanthropist from yet other angles. But what about an agro-ecologist—can she make head or tail of it all? In this chapter I shall try to do this by briefly pointing out all the problems of agriculture, and then arguing that the fundamental cause is a biological one. This is rational, since agriculture is after all a manipulator of biological systems. Indeed, there are also ethical, social, economic, political and aesthetic problems. But all these originate, I believe, in a lack of understanding of some very simple biological ideas.

Some people maintain that there is ample cultivable land to feed the growing human population of the world to the end of the century, and probably for long after (Hendricks, 1969). Nevertheless, there are two relevant factors which are very often ignored, or not considered at all:

1 Should every available hectare be used to produce the maximum amount of food for human consumption? This would inevitably be accompanied by large-scale landscape changes. To what extent should we plan and manipulate the biosphere wholly for the benefit of humans, perhaps at the expense of many other living things and the quality of our own lives?
2 Even if enough food is produced, and without further large-scale re-landscaping, there is no doubt that it will continue to be distributed unequally. For example, in 1985 enough food was produced to feed everyone, yet there was wholesale starvation in the Sudan and Ethiopia, and malnutrition in around three billion children worldwide.

It is said that problems with food production will be overcome by what is called the 'development of new technologies', or 'improvement in animal and plant varieties and general husbandry' (Marstrand and Pavitt, 1974). What are these improvements? Perhaps an entirely new approach is needed (IFOAM, 1986). If progress continues along the present lines, there is no reason to believe that these technologies will improve or that the present problem of modern high-input agriculture will be overcome.

There are four major characteristics of 'modern conventional agriculture':

1 It is an *industry*, requiring inputs many of which are applied in increasing amounts (Mellanby, 1975), unlike peasant agriculture which is more or less self-sustaining. These inputs, often

manufactured from inorganic sources, include fertilisers, pesticides, herbicides, fungicides and animal and human drugs, particularly antibiotics.

2 In the developed world at least, the agricultural enterprise is of a large size. Small units are, in general, vigorously discouraged. One of the major reasons for this is that, within the present economic climate (largely created by government grant aid), the larger units are considered more 'efficient' (but see p. 34).
3 These farms are highly capitalised, often over-capitalised; where large and expensive equipment could be used co-operatively, it usually is not but stands idle much of the time.
4 In conjunction with high capitalisation go low labour demands, in that the farms are low-labour demanding.

The consequences of these agricultural developments are varied. In the first place, because farming is regarded as an industry, then, as in any other industry, the prime reason for farming is to make money, not to grow food. It is curious that the undisputed necessity for food production is not seriously considered. Perhaps it would be more appropriate to consider food-producing agriculture as a service which, with its massive grant aid in Britain and the EC, it undoubtedly is (see p. 35).

Secondly, modern agriculture is considered by most agriculturalists and ecologists to be 'efficient' (Mellanby, 1975). Others are in trouble because they are not 'efficient' with their agriculture, and the best we can do is go and teach them to be efficient. Let us first examine what is meant by 'efficiency' in relation to agriculture. Are such statements true or false, meaningful or nonsensical?

The Myth of Agricultural Efficiency

The 'inefficient, traditional or non-modernised' small farmers of Europe are constantly being accused of living off the backs of the 'efficient' larger farmers, of Britain in particular, but also of other parts of Europe. This conjures up a picture of little dark men in berets, hand-milking their cows while sitting on rough-hewn milking stools, before pottering back to the vine-covered homestead to relax with a couple of cronies over a glass of home-trodden wine. Farming in this way is a *way of life*. It is *not* an industry. Certainly the brisk, modern, computerised nine-to-five industrial large-scale farm, with its battery of mechanical, chemical and medical aids, is an industry, but is it more 'efficient'? Should we aim at turning farming throughout the world into an industry, or should we consider that as a way of life it can often make more sense? Presumably we should make farming 'efficient', but what does this mean, and what do agriculturalists mean when they use such words?

Firstly, they point their finger at the large increase in gross yield of wheat, for example, which has risen from one ton per acre before the last war, to around three tons per acre or even more. Another example, often quoted, is the increasing yield of the dairy cow from 1,000 gallons per lactation to around 4,000 on average, and often more.

If one considers that 'efficiency' means an increase in *gross* production, then effectively this is true. But things are more complex than they seem. Can efficiency really be measured in terms of *gross* turnover? Indeed, we all know that businesses with very large turnovers can easily go broke; presumably this would be a sign of inefficiency, economically speaking at least. In economic terms efficiency means an increase in the profit margin. If we look at these agricultural enterprises, is there an increase in the profit margin? There may be, but this is not the full story. Until very recently, and in some areas of agriculture even today, the costs of this increase in production have been subsidised and grant-aided, either directly, as in the case of the production of fat lamb with a subsidy on every graded carcass (currently about £6 per beast), or indirectly, as in a grant on the construction of buildings (until four years ago as much as 40 per cent) or in the purchase of inputs such as fertilisers, pesticides, herbicides and machinery. Further indirect financial help comes from the vast sums that have been spent on agricultural research, financed out of the pocket of the tax-payer, which primarily benefits only about 1.5 per cent of the population who are farmers. The consumer has 'cheap' food, let us not forget, because it is financed out of his taxes, just as education is cheap if he goes to a state school (or in this case, 'free').

Other ways in which farming has been subsidised relate to its special status. For example, farms can write off any profit they make against 'improvement' by further capitalisation. This has led in particular to the ludicrous situation of a farm with one worker having three tractors! A few years ago a large Sussex farmer made so much money from corn and dairy that the only thing he could think of doing with the money to avoid paying tax was to make concrete roads along his farm tracks, which happened to be at the top of the chalk downs. Now the traditional footpath and bridle way along the top of the calcareous South Downs has concreted portions. The permeable chalk is really one of the last types of geological formation in which this is necessary.

It is common practice for people who have made a lot of money in other walks of life to invest in a farm primarily so that they can write off some of their other income against farm expenses. The farm can make a loss for seven years before the taxman queries it. This is another reason for the heavy capitalisation of British farms.

Overall the expenditure on agriculture each year is little less than that on education (HMSO Statistics on Government Spending). It is time we understood that present-day farming is neither primarily concerned with food production nor a money-making industry. Rather it is run as a *service*, using 80 per cent of one of the major resources of the country (the land) and employing approximately two per cent of the population. In this light it is arguable whether we are making the best use of this resource. The other 90-odd per cent of people are quite rightly beginning to question what actually goes on in agriculture (e.g. Carson, 1962; Harrison, 1964; Shoard, 1980).

Agriculture is not creating more jobs either. Hand in hand with the heavy capitalisation (which has been encouraged by the government subsidies, basically gifts) has gone the reduction of people employed directly on the land (from five per cent to 1.5 per cent of the population). It is far less difficult to buy a machine, often with some form of grant aid, than to employ a person and take on all the accompanying social responsibility. In fact another way the agriculturalists have of defining 'efficiency' is *the number of acres (or hectares) one man can look after*, with the help of course of a battery of mechanical and chemical aids. Thus efficient agriculture is the one where one man can farm 200 acres as opposed to 20. Effectively this means that efficient agriculture is where there are fewer people working *on* the land, and more *at* the factory bench making the equipment to keep that man on his large farm (Blaxter, 1976). This has led to the present situation in which it is considered impossible for one family to make a living off less than about 75 acres (31 hectares) of prime agricultural land unless they practise horticulture or intensive animal husbandry. The viability and importance of diversified small-holdings in providing food and employment have until very recently been ignored almost totally by the agricultural establishment.

Another area of agricultural aid is the guaranteeing of prices to the producer. Thirty years ago there may have been some sense in such policies, to encourage food production to feed Britain after the war. Nowadays it is difficult to follow the rationale of such extraordinary monopolies as the Milk Marketing Board who, it turned out, had the *right* to demand a payment from us when we were making and selling cheeses from our one cow! It is the real costs (instead of the cloaked costs, many of which are disguised by finance from the tax-payer, and by minimum fixed prices) that must be taken on board; the economic efficiency of agriculture is then questionable to say the least.

'Efficient', according to the *Oxford English Dictionary*, means *productive effect*. 'Efficiency' is *the ratio of useful work performed to the total energy expended*.

Ecologists are used to looking at biological efficiency in terms of the energy balance of organic systems, and if we are to do this with agricultural systems, we find that the modern high-input system shows a staggering energy deficit—one calorie produced to six used (Blaxter, 1975) and in the United States 1:10. When compared with a low self-sustaining system, the high-input industrial agricultural system looks deplorably *inefficient*.

It is important before assessing yields in terms of efficiency to consider the *real* costs (both economic and environmental). Studies in the corn belt of the United States as long ago as 1976 (Lockeretz *et al.*) indicate that in terms of energy and economics, modern high-input agriculture is very far from efficient. The environmental costs on local and international bases have still to be calculated and added to this.

If agriculturalists are seriously interested in 'efficiency', it would seem to be more rational to calculate the full equation and look at the *net yield* of the crop (gross yield minus the inputs and other 'costs'), whether this is done in terms of energy, economics, biomass or other indices. In later chapters we have done this for our three different farms (chapters 5, 6 and 7).

Efficiency of land use must also be considered. There are many people (the vast majority) who have other 'serious interests' in the land, which may be aesthetic, recreational or conservational. Modern farming strategies of 'industrialising' the countryside do not fit well with these, leading, for instance, to a lack of facilities for sports such as trial biking and horse riding, reduced woodland and hedgerows, draining of all wet areas, pollution of waterways and the homogenisation of the landscape.

What of the 'efficiency' of the animal industries on the modern farm? There are many facets to this, which are worth examining in some detail.

The rights and wrongs of raising animals in the types of environment that have been developed with a view to 'increasing the efficiency' of the enterprise have been heavily debated (e.g. Harrison, 1964; Singer, 1976; Reagan, 1982). In this case 'efficiency' may mean either the number of animals one man can look after, or the shortening of the lifetimes required to obtain the desired weight before slaughter, or both. Such 'efficiency' indices are open to serious discussion on many grounds apart from those of 'efficiency'. Let us, however, examine the latter. As René Dumont and Bernard Rosier accurately put it in the dedication to their book *The Hungry Future* (1969):

> To the children of backward countries who never
> attain their full promise,
> or who have died of kwashiorkor,
> because the fish meal [or soya beans, or grain] which
> might have saved them
> has fed the chickens [or cattle, or pigs, or horses]
> gorged by the rich.

If the animals, raised in large numbers at great cost in expensive, intensive, and until recently grant-aided buildings are to pay their way, they must grow fast, and to this end they are fed high-quality diets including often much protein food. These foods are frequently imported from Third World countries where the people have been persuaded to grow cash crops to sell. Because of the close proximity, large numbers and stressful conditions under which such animals are kept, both physical and psychological diseases are very common. Therefore rigid and almost routine drug treatments are needed. Indeed, drugs such as antibiotics are fed in manufactured foods in order to promote growth. This may be good for the drug companies, but it is difficult to justify on the grounds of efficiency—by any criteria.

This type of animal husbandry, which applies not only to veal calves, chickens and pigs (which have to date had much publicity), but also to dairy and beef cattle, ducks, geese, sheep and horses, is certainly not biologically efficient. Each species has evolved to live in its own niche and has adapted to its social and physical life to maximise its production in those conditions. That it has got it right is evident from the fact that it exists. A wiser approach to maximising efficiency or net yield would be to try and fulfil as nearly as possible those environmental conditions that the animal has evolved to require—at least *until there is good evidence to the contrary*. At the time of writing, this is lacking (see chapter 8).

The animal breeders will argue that, at least, by manipulating the breeding of plants and animals for use by humans, they have increased the efficiency of their animals. They have produced cattle that yield three times as much milk as 30 years ago, pigs that grow faster, chickens that lay more eggs, bigger calves at birth, that grow fatter (so big that the calves of some breeds, such as the Belgian Blue, can only be born by caesarian section). The efficiency of such developments should be very properly analysed or we are in danger of falling under the spell of the euphoria that grabbed the world after the plant breeders had produced the 'Green Revolution' (Dahlberg, 1979).

Let us take dairy cattle as an example. The 50 tons of milk per lifetime cow does indeed produce more milk, but at what cost, environmentally, economically, ethically, and finally to the cow herself?

In the first place, in order for this cow to produce this amount of milk, she must be carefully selected and usually line- or inbred by artificial insemination. This results in a reduction in the wealth of the gene pool. In the last decade we have, in Britain, seen the disappearance of at least seven breeds of cattle, and a continual expansion in the national herd of Friesians. Because of the small number of bulls used to cover the cows by artificial insemination (AI), they are rarely unrelated. The cow so produced responds well to the high levels of (often imported) high-protein concentrate food. Now the cows compete for food with humans, eating imported soya beans grown where there used to be equatorial forests in Brazil, perhaps with added fish meal (contributing to over-fishing). In relation to food resources, cattle evolved as a species that was complementary to humans rather than competitive, since they have a clever way of digesting cellulose with the help of a symbiotic relationship with another species—bacteria. Humans might well have a lesson to learn here about symbiosis!

Table 1. Comparison of the production and length of life of dairy cows on low input organic farms and conventional high input farms (from Kiley-Worthington, 1986)

Litres/Lactation herd av. 30 farms	*Numbers of lactations herd average*	*Butterfat %*	*Concentrates fed/ cow/day*
Low input; 5200	11.2	4.33	2.44 kg
Conventional; 6595	4.5	3.35	5.60 kg
Lifetime milk production/cow (Herd average)		Low input Conventional	51,464 23,400
% lifetime productive		Low input Conventional	79.9% 64.2%

Although the cow may produce a great deal of milk at each lactation, her life will be very short compared to that of a lower-producing cow who may stay in the herd for more than ten years (see Table 1). The reason for this is basically that she is stressed throughout her life and succumbs after only a few lactations (just over four is the national herd average now). As a result her lifetime yield is not as high as her neighbour's on the low-input system, and the percentage of her life during which she is productive is inevitably much lower. The high-yielding, high-protein-fed, housed dairy cow, in large herds, tends to be prone to ill health and requires a very high standard of stockmanship (which is not always available) coupled with a battery of drugs. The prevalence of sub-clinical mastitis in large herds, for example, is extremely high, despite a high level of cleanliness and apparent 'good husbandry'.

The cow will have her calf taken away from her shortly after birth, although she is the best at looking after and feeding it. The calf will be fed a milk substitute from the milk lake, made from its mother's milk which has been transported often from one side of the country to the other and back, processed, and returned to the farm. This is mixed up in buckets and fed back to the calf. Raising calves away from mothers in this way leads to many physical and behavioural problems which, even with the 'best' husbandry, can result in relatively high levels of mortality (around two per cent is fairly normal). The cow will be deprived of sexual experience since she is unlikely to have access to a bull; rather she will be impregnated by artificial insemination when she comes into oestrus. This in turn is less efficient at causing pregnancy than a bull. A spokesman for the Milk Marketing Board indicated that the cost of artificial insemination to the national herd was in the region of £50 per cow per year (as a result of loss in yield because of failed conceptions, inability of management to recognise oestrus, or cows not showing oestrus). When asked why farmers did not then have a bull, he replied, 'This is what we are afraid might happen, they might find it cheaper' (Esslemont, 1984). In the light of such evidence one tends to feel that the most *efficient* arrangement might be to leave it to the bull!

Some ten years ago, the Ministry of Agriculture, worrying about the over-production of dairy products, offered farmers financial incentives to have their entire herds slaughtered in order to go out of milk. Many herds, some containing some of the best of minority breeds, were slaughtered, but ten years later we *still* have a growing milk lake, butter and cheese mountain and yogurt swamp. Why? Because farmers have decreased their number of cows, but by increasing feeding, drug use and the artificial environment for cows have managed to increase the yield per lactation per cow, although not the number of lactations. They have produced *more milk at greater all-round cost*. Is this an increase in *real* efficiency? An alternative would have been to give incentives for farmers to decrease their inputs and diversify their farming activities, including keeping some suitable dairy cows on a low-input system. Thus they could have increased the *net* yield from each cow while decreasing the overall gross yields, and cutting environmental, aesthetic and ethical costs, as well as economic . . . but they didn't think of it, or if they did there were too many powerful vested interests in the agricultural lobby to allow this to be considered seriously.

Another particularly inappropriate and unjust development has been that of milk quotas, which were brought in when the financial incentives to slaughter milk cows had not worked well enough to make a real difference to milk production. Those still producing milk were given a quota and told they could only produce that amount. But if you give up producing milk you do not have to return your quota to the government, which would have been fair, as it could subsequently have been either withdrawn or reallocated. Oh no, you are allowed to sell it! The result is that farmers who were producing milk in quantity when the quotas came in, when they retire have not only their farms to sell, but in some cases can obtain *more* money from selling the milk quota. So the rich have become very much richer, often millionaires, and the over-production of milk continues at considerable environmental, social, economic, ethical and aesthetic costs to the country. Is this efficient agriculture or just how-to-get-very-rich-at-much-cost-to-others agriculture?

There are agricultural alternatives, ideas and methods that are based on low-input, self-sustaining, more net-efficient systems. The International Federation of Organic Agricultural Movements held a three-day meeting in California in 1986 to discuss research development and the new agricultural techniques (IFOAM, 1986). The British government has been almost unique in ignoring such research. It finances hardly one research assistant working on these problems; there are no degrees, no lectureships even, in such subjects. It is even impossible to have a youth employment training programme teaching alternative agriculture: it must all come under the umbrella of the conventional agricultural establishment, the local agricultural college. I know because I have tried.

Thus almost all the theoretical and practical research on alternative low-input, sustainable, more efficient farming systems is being done without any financial aid from the government.

With growing unemployment, increasing costs of inputs and fluctuating land prices, it would seem important to examine the real efficiency of agricultural strategy. The world is growing short of many resources, and starvation is still rampant, although in the West we have surfeits of expensive foods produced often with the help of resources from the starving regions of the world. Perhaps agriculture is becoming too serious a subject now to be left only to the agriculturalists, or even to big business. Perhaps the small farmer in his beret, smiling over his wine, has something. Maybe his farming is not merely a way of life, but also more 'efficient': *a higher rate of useful work performed to the total energy expended.*

The 'inclusive costs' of modern agriculture can be divided into biological, social and political, aesthetic and ethical as well as economic costs. A fair assessment of the problems locally and globally must examine all of these.

Biological Costs

Since agriculture is fundamentally about growing animals and plants, by definition it is involved with organic, biological systems. The basic problem of industrialised agriculture is that the operators and researchers lack understanding of how these systems work, and from this spring all the other problems. The two most important central concepts are:

1. The breakdown of the self-sustaining ecosystem.
2. The reduced diversity of animal and plant species on the unit.

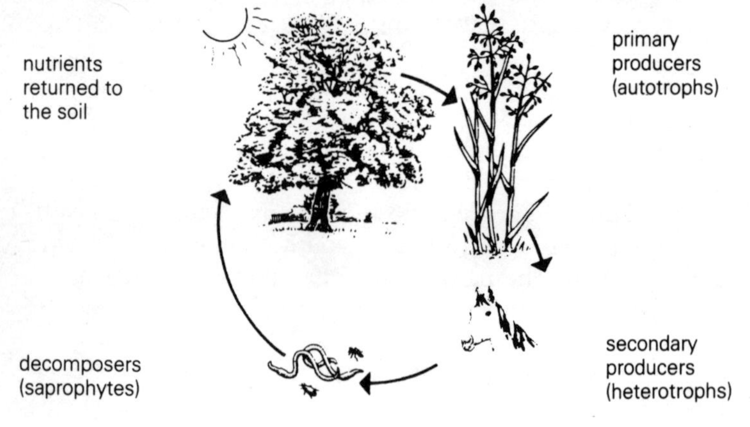

Figure 1. The self-sustaining characteristic of ecosystems.

Firstly, any natural biological system is *self-sustaining*. Thus the sun's energy is fixed by the autotrophs (primary producers such as green plants) which feed the heterotrophs (including secondary and tertiary consumers such as pests, herbivores and man). The waste products from either or all of these groups are then broken down by decomposers (also called saprophytes) and the minerals and nutrients so liberated are taken up by the next generation of autotrophs, so giving life to the heterotrophs (e.g. Odum, 1971. Figure 1). There are huge numbers of feeding cycles within cycles. The logo for our farms represents one simple ecological feeding

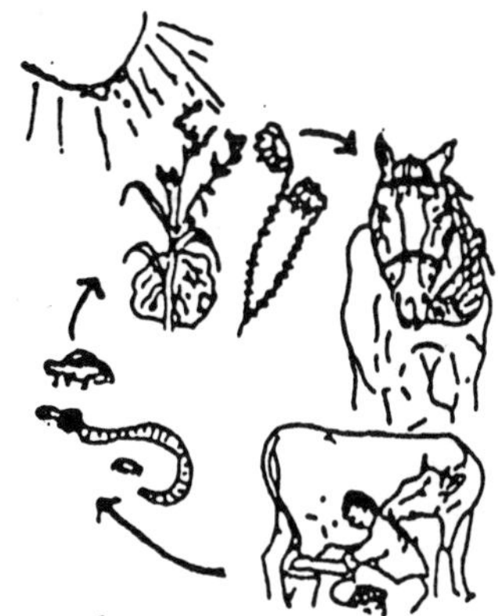

Figure 2. The Eco-Farms logo.

and waste return cycle (Figure 2). During this process, humus is added to the soil which generates a suitable environment for saprophytes, and good growing conditions for autotrophs. Very few minerals or nutrients disappear and are lost to the system, they simply go round and round in various small and large cycles powered by the sun. Thus the system is self-sustaining: it will continue in much the same way, with the same animal and plant species balanced, until or unless there is some major ecological disturbance.

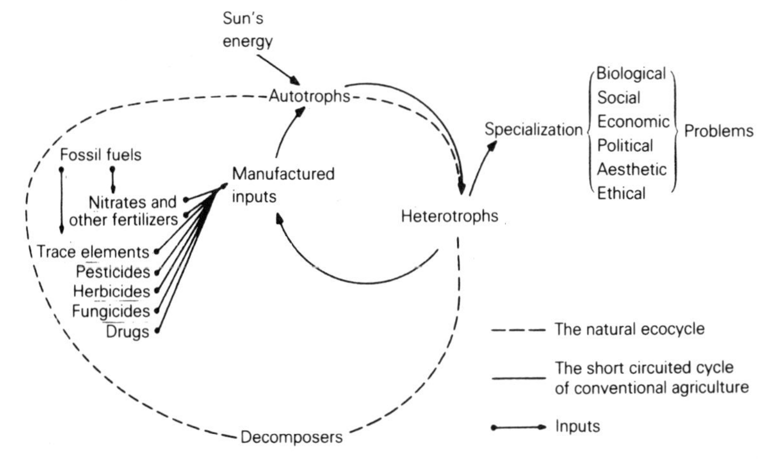

Figure 3. The breakdown of the natural ecocycle in modern agricultural systems.

Modern agricultural developments have broken this cycle by reducing or eliminating the decomposers as a result of change in the soil environment, which has been created and is maintained by high inputs (Figure 3). In modern agriculture there is often no substitution for the loss of humus. Humus acts as a buffer for extremes in soil conditions, lightening heavy soils and retaining moisture in light soils, as well as providing material for the decomposers to feed on. As the amount of humus declines, so do the numbers and types of decomposers, making the system dependent on ever-increasing inputs to supply nutrients for plant growth, rather than recycling nutrients that could be released from the soil structure. Thus the natural fertility is reduced, and the natural biomass declines. This process is not irreversible, but once the soil has become impoverished in this way it takes a long time for reconstruction to take place.

It is often argued that this is not necessarily a bad thing; if yields are high and it is profitable, such a system is 'better' than one which is self-sustaining. The problem is that this is not the only effect of such farming techniques.

Another biological problem is that of *reduced diversification*. This is heavily encouraged by the current agricultural establishment. Specialisation is said to make more money, be simpler to operate and reduce labour required. However, a simple rule in ecology is that the more homogeneous the system (the less species diversity there is), the less stable it is, and thus the more vulnerable to disaster (Leopold, 1947; Odum, 1971). There are some exceptions to this rule but it is generally the case. Therefore, one of the results of the encouragement of monoculture and specialisation in agriculture is reduced stability. Unstable systems are more prone to disease attack, and pests. The agriculturalists' response, so far, has not been to return to a more diversified husbandry, but rather to use chemical controls in the form of pesticides, fungicides, herbicides and drugs. The effect of the wide-scale use of pesticides and their side-effects has been well documented (Carson, 1962).

The abuse of antibiotics by over-use in animal husbandry has received less attention (Swann, 1969). Antibiotics have been used as growth promoters or for prophylactic purposes in animal feeds, resulting in the development of bacteria resistance and residues in the meat.

Since the life-cycles of all species within the ecosystem are interlinked, interfering with certain populations (of pests, for example) can cause unforeseen results, such as the reduction or elimination of their predators. This in turn ensures that natural controls are less effective, and the system becomes more reliant on applications from outside.

In some parts of the world the natural fertility is further reduced by *soil compaction* (e.g. the clay soils of Midland England and the Tropics). Where arable soils are heavy, the frequent passing of heavy machinery over the area results in reduced aeration of the soil, reduced numbers of decomposers and hence breakdown of humus and liberation of nutrients, reduced drainage and growing potential. In lateritic tropical soils it also encourages the formation of a 'hardpan' (impermeable layer of minerals below the topsoil) which can increase run-off and erosion (White, 1981).

A further effect of reduced diversification and increased inputs is *agro-industry pollution*. This, on a world scale, pollutes waterways, river systems, soil, sea and air and causes large population changes in wildlife, which may, in turn, affect agricultural production. Well-known examples of agro-industry pollution include the accumulation of DDT in birds' eggshells, and the increased algal growth in waterways as a result of nitrate run-off. Some cases are more surprising, such as dangerously high levels of nitrates in municipal water as a result of run-off in Norfolk at the end of the drought in 1976. At this time mothers with small babies were advised against using tap-water, and were issued with special low-nitrate water.

In order to farm in the modern conventional way *resources are needed from outside the farm* and indeed often from outside the country. For example, Britain is reliant on Moroccan phosphates for fertilisers. There is no reason for phosphates to be imported: it is relatively easy to recycle phosphates if human and animal sewage and bone remains are treated appropriately and returned to the soil, as is in fact done in parts of China. Rural district councils in parts of the United Kingdom (e.g. Wealden District Council, Sussex) offer free delivery of raw or treated sewage which has been tested for toxic chemicals, but few farmers make use of this service. Recycling human sewage from cities would be more difficult, although there would need to be stringent controls on industrial wastes and toxic components.

Britain is dependent on another scarce non-renewable resource:

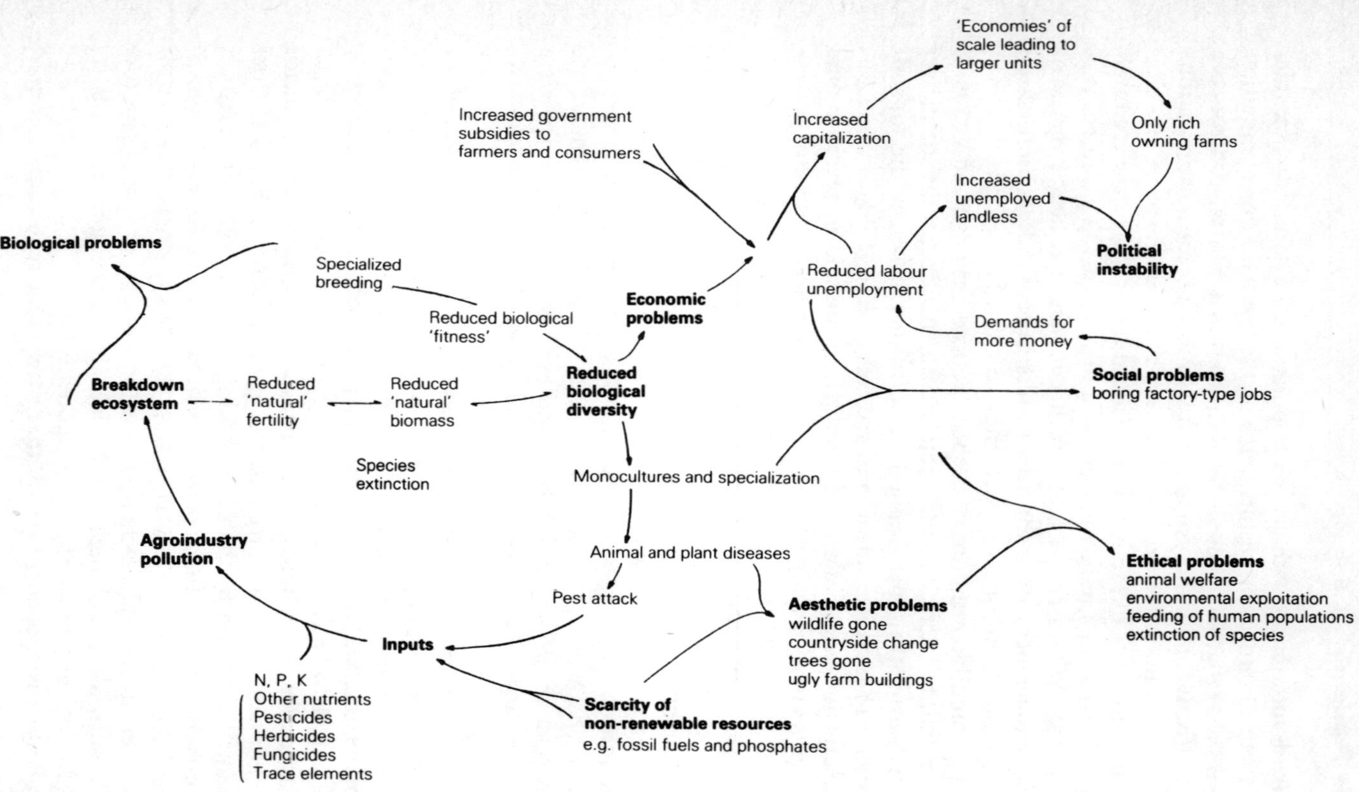

Figure 4. Problems of modern high input agriculture.

the use of fossil fuels in agriculture. Although only eight per cent of the fuel used by the nation is used on farms, nevertheless this is still a substantial quantity, and much more is used in the transport and processing of foods (Blaxter, 1975). Renewable sources of energy, such as solar, wind and water power, will be more effective in the future, but until they are easily available we shall continue with current agricultural policy—unless research on renewable energy sources becomes better financed, which is unlikely at present (it accounts for around one per cent of the energy research budget, while nuclear energy accounts for 80 per cent).

The biological and other problems are summarised in Figure 4.

Social and Political Costs

Specialisation in agriculture has meant that there must be specialisation of jobs on the farm. Today the farm hand is designated as the 'pig man', 'tractor driver', 'dairy man' or 'calf rearer'. He does the same job continuously on the farm, where he used to do many; in many ways farm work is being converted into a factory-type, repetitive job. Although the working conditions may be easier than before the war, they are often duller, and the added hazards of large amounts of agricultural chemicals and big and dangerous equipment do not improve the situation. Since the farm worker is doing a factory-type job, he or she also demands a factory-type wage to compensate for low job satisfaction.

The farmer's response is to capitalise further and reduce the numbers of farm workers. Consequently there are more landless and unemployed in both Europe and the developing countries. In the latter case there are not the welfare payments available to keep them in the rural areas, and they drift to the cities where they are confronted by the same problems of greater mechanisation, reduced labour demands and more unemployment. Many who are without land are unable to sustain themselves. In Europe, with higher capitalisation and rising land prices, only the rich can farm and the rural farm worker finds himself unemployed. Here he does not starve but lives on social security payments. In less favoured areas such as the Scottish Highlands and Islands, the farmer himself lives on subsidies (see chapter 6).

The presence of large numbers of unemployed, and the fact that only the rich are able to own land, leads to a politically unstable situation, particularly where there are increasing demands from people to be able to get back to the land. In the United Kingdom these people are now forming political movements, and in many other countries they are likely to become a force that governments must consider.

Aesthetic Costs

The original conservationists' arguments were based on aesthetic grounds (Tansley, 1945; Leopold, 1947): 'it is a pity to destroy all the wealth of wildlife we have around because it is unique and beautiful.' However, there are also strong economic arguments for the preservation of wildlife. For example, elimination of small predatory animals such as stoats, weasels and birds of prey leads to increases in populations of game birds, but also of pest birds such as pigeons and starlings. Similarly, insecticides not only pollute areas and become deposited in birds' eggs, they also kill pests and their predators. Large-scale interference with the ecology of a region may have wider implications as well as aesthetic effects (Carson, 1962).

The removal of woodland and hedgerows, the drainage of marshes and the enlargement of fields not only change the face of the countryside, but also result in substantial changes in wildlife populations and the possibility of the extinction of some species (Shoard, 1980).

Conservationists usually deplore developments in agriculture which destroy wildlife, but remain convinced by the agriculturalists' arguments that their techniques are 'efficient' and vital for food production. Thus they have tended to concentrate on the establishment and maintenance of nature reserves rather than the wider issue of the impact of modern agricultural developments on the countryside as a whole.

One development which has received less attention, and yet causes an ever-increasing scar on the countryside, is the construction of low-cost agricultural buildings alongside traditional buildings of local materials. Until 1974 planning permission in Britain was not required for such buildings and still is not in many countries. Now planning permission is only required for buildings in excess of 1,000 square metres. But for any form of human housing there have been much more stringent controls for more than three decades. If you wish to put a window in your house, or erect a garage, you must have permission. My brushes with the planning authorities have pointed out some extraordinary stipulations. Did you know it is illegal for you to curl up in the hay and sleep in your barn if you wish? Take care, all ye Scouts, you might be run in on your expeditions!

While planning control is essential in a heavily populated country like the United Kingdom, the degree of discrepancy in the controls over human habitations and agricultural buildings makes little sense. We should perhaps also consider personal liberties—provided education is appropriate so that they do not cause large-scale environmental problems.

Ethical Costs

There are three main groups of ethical problems associated with modern agriculture:

1 If we believe that it is a 'human right' for everyone to have enough to eat, then how is it that although enough is produced in the world, the number of starving is increasing? Clearly, modern high-input agricultural strategy is not working in terms of feeding the world. Is it right, therefore, that it should continue in this way? (See chapter 1.)

2 If we believe that it is a 'human right' that the next generation should inherit a world at least as habitable as the one their parents had, then we must very seriously understand the environmental effects worldwide of this modern 'money first' agriculture: the destruction of the Amazonian forests to raise cash crops for the United States; the destruction of the forests of South East Asia to raise money for the rich; the pollution of the Norfolk Broads in Britain by agricultural run-off; the destruction of wildlife and subsequent disturbance of the ecology of each area worldwide.

3 If we believe that there are other sentient beings on the earth and that they should at least have a life free of prolonged suffering, then we must consider how, where, when and why we have animals, and in particular the intensive management of farm animals (chapters 8 and 9 discuss these questions in detail).

I have discussed the problems inherent today in high-input, large farming practice which is being encouraged by governments throughout the so-called 'developed' world, and more seriously is being established in the 'developing countries'. The training of bright young agriculturalists from Third World countries in conventional agricultural techniques, which they then apply on returning home, can have quite disastrous consequences. For example, starvation occurred in Guatemala in 1972, as a result of the decline in subsistence crops in favour of banana production and the subsequent market collapse. It would seem essential that if Third World agriculturalists are to be trained in other countries then *they must have alternative agricultural practices suggested and presented to them*. It is important, too, that the women who may be practising agriculture have access to this information and cultural exchange (Chapter 10).

Similarly, it is equally important that agricultural advisers in developing countries should temper their ideas of high-input agriculture with alternatives that have less traumatic biological, social, political, aesthetic and ethical effects. Since they have been trained to think quite differently (because there have been no other courses available) it is highly unlikely that they can do this at present. Recently one or two agricultural colleges in Britain have started offering courses in 'development agriculture' or 'sustainable agriculture', but still taught by the same staff. This may be a step in the right direction but we must move fast if the problems are not to outstrip the evolution of our agricultural thinking and teaching.

Enough of criticism and destruction. Is there an alternative? Could it solve at least the majority of these problems, and most important of all, could it work? This is what the rest of this book is about: *Ecological Agriculture, Food First Farming*.

3 Ecological Agriculture, What is It?

I am aware that movements promoting alternative agriculture already exist, and have done since the 1920s when Howard in India, for example, thought about the possibility of using waste materials more sensibly. His ideas were subsequently built on by Lady Eve Balfour and several others of the stalwart 'muck and magic' brigade.

This group called themselves Organic Farmers. Then Rudolf Steiner thought about agriculture and published his books on what has come to be called Biodynamic Agriculture. Most recently, Bill Morrison in Australia coined the term and the ideas of Permaculture. Each of these approaches has something in common with Ecological Agriculture, but they also differ, and I must explain why we need yet another name by pointing out the problems with these other approaches.

Organic Agriculture

The ideas of the early 'muck and magic' brigade in Britain developed primarily as a result of worries concerning the declining quality of the diet due to eating food grown by modern high-input agriculture, and the consequent reduction in natural soil fertility (Balfour, 1975). However, they also often had a strong spiritual or intuitive base for their beliefs and ideas. Very few of these early pioneers were scientists but they were a determined small bunch with great courage and conviction.

In the 1970s their doctrines began to be considered and sometimes even embraced by the 'Flower Power' generation. They rallied anew, and appointed E. F. Schumacher, a like-thinking independent (but a clever man who had managed to maintain his position within the establishment and therefore was well listened to), as president of the Soil Association. This group used the term 'Organic Agriculture'; the French equivalent was *Agriculture Biologique*. Since agriculture by its very nature is both organic and biological, these terms have been the source of much contention and misrepresentation through the years.

No one has bothered to define Organic Agriculture. The result is that when one does try to find out exactly what people mean by it, they all seem to interpret it differently—from the woman in her sandals and denim skirt with her cow and spinning wheel, who sees it as a semi-spiritual type of tilling of the earth in harmony with earthly rhythms and menstrual cycles; to the large-scale arable farmer in his tweeds, green wellingtons and stuffed machine-washable waistcoat, whose daughter hunts with the Quorn and who produces 'organic' wheat to sell at a premium to supermarkets. The only difference in this man's practices is that he has bought in crushed and dried seaweed, or manure from his neighbour's piggery, instead of nitrogen fixed in a factory from the air and sold in a bag. He may also have put on basaltic ground-up rock or rock phosphate instead of buying it in a bag after it has been through the hands of a multi-national chemical company. The hill sheep farmer will also maintain that he is an 'organic' farmer because he does nothing to his land at all except let the sheep graze it and in their wake create a wet desert.

The only factor 'organic' farmers seem to have in common is that they do not use bagged nitrogen fertiliser. Such a state of affairs, with such different, and sometimes rather muddled thinking, did not convince me that here lay the answer to the problems of modern agriculture.

Although organic farmers may reduce some of the problems of high-input agriculture, they could also create others. For example, it defeats the whole object if the farm is dependent for its functioning on the manure from an intensive pig unit (a 'processed animal protein' recently marketed by organic farmers and growers) run along exactly the same lines as any normal high-input pig unit. This involves buying in of high-protein food (often imported from Third World countries) for the pigs, and housing them in intensive systems where they are subject to considerable behavioural deprivation and which are open to serious criticisms from those interested in animal welfare. The result of such practices might well be an increase in intensive animal housing systems in order to provide the muck for the 'organically' grown crops consumed by humans. This, I would suggest, would be a move towards an even less acceptable agricultural system than the present high-input, bagged nitrogen strategy.

It is equally suspect for 'organic' farmers to become dependent on dried and processed seaweed to provide nutrients. I have not yet seen any serious criticism or even thought-provoking discussion on the acceptability of this for 'organic producers'. No one has paused to think of the effect on the seabed of a growing demand for large amounts of seaweed. Certainly we must expect large-scale changes in the fish and invertebrate populations as a result of changing the ecology in this way. There might indeed be a case for cropping in a sustainable way certain portions of the sea floor for seaweeds, but until we have much more ecological information on the environmental effects of growth and production of seaweeds, it would seem rash, unacceptable, to do it willy nilly—and certainly not environmentally sensible.

I well remember, some ten years ago, when it had been suggested that we should run some courses in 'organic' gardening at our farm, being 'inspected' by a well-known lecturer in 'organic gardening'. Apart from suggesting that we should give two-to-three-hour lectures, which he then proceeded to give me (no one can sit this long, however well motivated!), he then became outraged because we did not use peat in our garden. Since we were on alluvium in a downland valley, there was none on the farm, and this would have meant buying it in, probably from the Somerset Levels. It so happened that a couple of nights before there had been a television investigation of the Somerset Levels. Among other things, this had pointed out the degree to which this unique habitat was threatened by the steadily rising price of peat as a result of its great popularity with gardeners. This lecturer had apparently not considered where the peat came from, or that he might be contributing to the loss of yet another habitat with his constant reiteration of the glories of using peat in 'organic' gardens. It would have been a little more acceptable if he had at least *thought* of this, and had produced arguments to justify using peat anyway. To be fair, there is now concern about the dramatic increase in the amount of peat extracted for gardeners—but this as usual is the result of public pressure. It has not come from the organic advisers.

So until the term 'Organic Agriculture' is better defined, and its proponents have thought seriously about all its implications, it might be better to avoid its use and the risk of confusion.

Biodynamic Agriculture

Followers of Rudolf Steiner, by contrast, have a well-defined alternative agriculture. This is called Biodynamic Agriculture (Koepf *et al.*, 1976). Steiner was an extremely prolific writer. Apart from his better-known writing on education, which gave rise to the Waldorf schools and Camphill communities that perform a tremendous social service, he devoted at least one entire book to his ideas and practical instructions on agriculture. Biodynamic agriculture is now well established, particularly in Europe, but there are anthroposophists (followers of Steiner) throughout the world.

Anthroposophists are Christians who believe in man's need to steward nature. According to their faith humans are completely separate from other animals; indeed, Steiner was also a sexist, stating that human males should have certain roles and functions on a day-to-day basis and human females others. Although anthroposophists are often 'good' and peaceable people, and have a real spiritual appreciation of the world, nevertheless their approach is basically anthropocentric. The world exists at least in part for humans, who were made in the image of God. It is a marvellous place and must be 'stewarded' by benevolent people, in order to maximise its benefits for present and future humans.

Biodynamic agriculture takes heed of the natural rhythms of the stars and moon. One of the practical spin-offs of this is that different plants must be planted at different times of the month depending on whether they are a root, a stem, a fruit or a flower crop. These dates are worked out and printed in a calendar and are based on the interaction of the elements of earth, fire, water and air for that type of plant at any time. Although, to a scientist, some of their practices and beliefs seem somewhat extreme, nevertheless some of the first and most impressive pioneers of an alternative agriculture were anthroposophists.

It is largely in the field of livestock management and production that I find major irrationalities and difficulties with their approach. For example, they believe that cattle should not be dehorned as they are then not 'complete'. This is a view with which one can sympathise: surgery on livestock or humans (whether it is debeaking, dehorning or castrating) should not be carried out as a matter of course; preferably it should be totally avoided, or at least be subject to careful consideration (see p. 176 *et seq.*). However, I was surprised to find that although they would not dehorn their cows, the biodynamic farmers were not averse to yoking their cattle for the entire winter in narrow standings. If we were to ask the cows if they would prefer to have their horns removed when they were calves, or be chained up for a very large portion of their adult lives, my guess (and my professional opinion as a result of studying cows' behaviour in depth) is that they would prefer the former. Yoking for prolonged periods, although not exactly illegal in Europe yet, is against the codes of practice on farm animal welfare (HMSO, 1980). But if the animals are horned and kept in very restricted quarters, it is necessary to prevent them injuring each other. Biodynamic farmers do not seem unduly concerned with such apparent irrationalities.

Another example comes from their use of herbal remedies. These are so well documented and established that there are factories actually making them, and the preparations they use for compost making. Certainly the excessive use of manufactured medicaments is to be avoided in both human and animal medicine. There is little doubt that herbal remedies can be extremely efficacious and I would be the first to applaud further research and development in this field. But the anthroposophists can be somewhat doctrinaire, and often do not use, for example, antibiotics at all on themselves or their animals. Mastitis is treated with garlic. Although one is entitled to make one's own decisions about what one does to one's own body, nevertheless the morality of making decisions for one's animals, which will inevitably cause them more pain and suffering, is disputable. I have visited several in other respects very well run and impressive biodynamic farms, where a substantial portion of the dairy cattle have only three working quarters to their udders as a result of not using antibiotics to control acute mastitis infection. Another problem that can become evident in the stock on biodynamic farms is chronically high worm burdens, which can easily be controlled by anthemintics (worm-killing medicines), but which are not used.

Thus although the Biodynamic farming movement is well established and there certainly are some well-run, productive farms which go a long way to solving many of the problems of modern high-input agriculture, there are areas within their philosophy, theory and practice which I find difficult to absorb. There is little empirical evidence for many of their beliefs. Although, perhaps, one should not be unduly thrown by this, nevertheless their philosophical position I find unacceptable. The thinking through of some of their systems, in particular animal management, is incomplete.

'Sustainable' Agriculture

One of the fashionable agricultural alternative terms is 'Sustainable Agriculture'. The International Federation of Organic Agricultural Movements (IFOAM) and others have had various conferences on this. 'Sustainable Agriculture', although in many ways an enormous improvement in thinking, is still, it seems, incomplete. Sustainable at what cost? Is it all right to continue to use chemical inputs so long as they do not have detrimental effects on the soil? Can animal foodstuffs be imported from the other side of the world so that animals can be kept to provide manure for soil improvement over a prolonged period? Is tree-planting in arid areas sufficient to prevent wind erosion and decrease soil salinity (Campbell, 1991)? There is little doubt that the biosphere is a sustainable system in any case. Indeed, according to Lovelock (1987), environmental changes caused by human activity have had remarkably little effect to date. But is this enough? Do we not have to think about all the costs emphasised in the last chapter? In a sustainable system, what will it be acceptable to bring onto the farm and in what circumstances? Where do we draw the line for 'sustainability'—at the farm gate, the shire, country, continent or world level? It may vary, but inevitably, to ensure sustainability everywhere without costs to others we must think about these boundaries. Many aspects of 'Sustainable' Agriculture remain to be thought through and defined.

Permaculture

Another approach which is gaining ground among thinking would-be farmers and gardeners is 'Permaculture'. This approach was thought out by Bill Morrison (1988), a plant ecologist, at about the same time as we were working out Ecological Agriculture. It is also based on involving ecological ideas in plant growth and human living design. It encourages, for example, the use of microclimates, three-dimensional use of the garden or window box, reducing inputs and so on. His book is full of interesting, innovative ideas and techniques for growing food and living well. In many ways his approach resembles Ecological Agriculture, but it concentrates on plant growth and human living. The thinking concerning other animals and how they should live and be treated, how to fit them into the systems and live symbiotically with them, is again, I feel, not his strongest point and needs further consideration.

A combination of techniques and of thinking through the ideas to produce a workable synopsis for the future is, I hope, what 'Ecological Agriculture' or 'Eco-Culture' can achieve.

Philosophically, in a nutshell, it is concerned that:

1 sentient beings should have more equal consideration.
2 the producing of food and farming in this way should not cause environmental, biological, economic, social, political, aesthetic and ethical problems such as confront conventional agriculture.

It is an attempt to marry the approach and knowledge of the ecologist, ethologist and philosopher with that of the agriculturalist, and to mix it up with the amateur philosophy of a surviving female peasant. This chapter outlines how Ecological Agriculture might be achieved, in particular identifying its central tenets.

First, let us be quite clear that we mean by Ecological Agriculture. It is an attempt to use our knowledge of how ecosystems work and apply this to agriculture in order to try and increase the biological efficiency and reduce environmental problems. There are two characteristics of ecosystems that must be understood from the outset:

1 *The interrelatedness of living things with each other and the rest of the environment*

Wiping out a single species, destroying even one parasite, will affect other living things and the environment around it. Whether we like it or not, particularly dramatic man-made environmental changes will have other environmental effects. Not all of these will be destructive or a nuisance to humans, but it is as well to understand that there may be many unforeseen consequences. If we do not know and cannot work out what these will be, then we should proceed cautiously until we do.

We should also understand that although there is interrelatedness of all living things in the biosphere, nevertheless the biosphere itself is made up of relatively discrete ecosystems—for example, the pond at the bottom of the garden (a favourite for school and university ecology classes); the Sahara desert; the Amazonian forest, and so on. The boundaries of each of these ecosystems may not be very obvious. Where does one end and the other begin? One can study how the ecosystem of a rotting log operates (another favourite for students)—the species that live on it, their food habits, population dynamics and how the nutrients circulate. Alternatively, one can study the ecology of the forest where the log is found, or the island where the forest grows, and so on. Thus although all things within the biosphere are to some degree mutually dependent, there are within this biosphere many ecosystems.

One might define an ecosystem as '*a place where the mutually dependent relationships between the species are stronger than those between ecosystems*'.

2 *The self-sustaining nature of ecosystems*

The dependency relationships within ecosystems are crucial. The log can go on rotting long after the wood has burnt down. This is because one of the fundamental characteristics of ecosystems, which also helps to define them, is that they are to a large degree *self-sustaining*. Understanding this is crucial to developing ecologically sound agriculture.

How, then, is ECOLOGICAL AGRICULTURE defined? It is:
***the establishment and maintenance of an ecologically self-sustaining low-input, economically viable, small farming system managed to maximise net production without causing large or long-term changes to the environment, or being ethically or aesthetically unacceptable*.**

To achieve this, there are certain tenets:

1 It Must be Self-sustaining, Including in Energy

Compared to agricultural systems, natural ecosystems are relatively self-sustaining. However, there are differences between nutrients and other elements in the extent to which this is the case. There were some famous experiments done by ecologists in the 1960s in the United States. They measured all the incomings and outgoings from a natural forest. These were called the Hubbard Brook experiments. Even in this steeply sloping pine forest the nutrient losses were surprisingly small and some important nutrients actually increased (e.g. nitrogen). Since then, what is happening to the rain forests in the Amazon is bringing home to the layman how efficient and remarkable the natural ecosystems can be—and how vulnerable. On poor soils, the most biologically productive, diversified and abundant forests have established themselves over centuries from the resources that were there. Cut down the forest and all those centuries of accumulated life-giving resources can be destroyed in less than a decade.

The ecological farm must be managed to create life-enhancing characteristics of the natural ecosystem, not to ignore or destroy them as has been the predominant approach of conventional agriculture.

What, then, are the most important relevant factors in creating a self-sustaining, prolific, diversified and abundant agricultural ecosystem, and how can it be done? These can be broadly placed in three categories:

a Nitrogen levels can be raised and maintained by the use of plants that live symbiotically with nitrogen-fixing bacteria (e.g. clovers and other members of the Papillionacae, the pea family), blue-green algae, and the creation of conditions appropriate for their growth.

b Recycling of nutrients, trace elements and minerals can be facilitated by encouraging decomposers, principally by creating good living conditions for them. One of the most important factors here is food supply: plenty of humus and organic wastes of one sort and another. However, Ecological Agriculture does not involve the importing of organic wastes or surfeits from long distances away, such as peat or straw bought from the other side of the country, or seaweed extracts. The widespread application of such products will inevitably cause further ecological problems such as is happening with the extraction of peat from the Somerset Levels in Britain. Animal manure imported from other animal enterprises is not acceptable either. Rotations, green manuring (growing plants to be ploughed in) and appropriate management and encouragement of species-rich pastures can aid the achievement of the productive self-sustaining system. For example, deep-rooting plants such as docks can help bring up minerals that are in the process of being leached out of the soil. Different plants and even different parts of different plants can concentrate a variety of minerals, nutrients and trace elements (Holder, 1978). If we had enough knowledge here in principle we could design the mixture of species in the grazing sward so that losses would be almost negligible.

Occasional application of nutrients or minerals found to be lacking when such a farming system is begun, or to correct pH (soil acidity measure), can be used to reconstruct soil fertility, but thereafter correct management should be sufficient to maintain and improve it, except in very extreme cases.

c Any minerals and nutrients lost off the system by sales (of plants or animals) must be replaced not by the application of large amounts of manufactured inputs, or even natural minerals from other areas, but rather by recycling (such as, for phosphate, the application of bone meal and sewage from the animals and humans supported on the farm).

To be a self-sustaining system, the greater part of the food for the animals and people on the farm should be produced on it. In most areas of the world there are non-growing seasons for which food must be conserved during the growing season (winter in temperate areas, and dry seasons and drought in tropical and sub-tropical areas). It is remarkable how little this is habitually done in tropical and sub-tropical parts of Australia, Africa and South America.

In addition, the system should not consume large amounts of energy. A natural ecosystem is powered by the sun (see Figure 1, p. 42). Even then the plants only use about ten per cent of the sun's energy. By contrast, British high-input agriculture consumes six times as much energy as it produces (Blaxter, 1975). In the United States it can be worse, as much as ten times (Steinhart and Steinhart, 1974). The conversion of an energy-balanced system, which all natural biological systems have to be, into an energy-consuming system of some magnitude illustrates perhaps better than any other aspect the lack of biological efficiency of the high-input agricultural system (see p. 36).

2 It Must be Diversified

In a natural ecosystem there are different 'niches', sometimes with slightly different climates, amounts of light, soil structure, shelter and so on, which are occupied by different species of plants, and as a result different animals that live with and off them. Most ecosystems are therefore extremely diverse. A species-diverse ecosystem has much more chance of surviving if there is a drought, a flood, or other natural disaster or change, since there are likely to be at least some plants and animals within it that can adapt to the new conditions. Generally, therefore, diverse systems not only facilitate self-sustainability, but also encourage stability. It is the 'don't put all your eggs in one basket' adage (Odum, 1971; Dorst, 1971).

Thus, if a farm is to be self-sustaining it needs to be diversified (having both animals and plants and different species of each). Different habitats and terrains will be required on the farm to accommodate these different plants and animals and provide a veritable plethora of niches for the appropriate species to occupy.

3 The Net Yield Per Unit Area Must be High

When it reaches its climax, the natural ecosystem will produce the maximum amount of biomass (biologically produced material) for that area at that time in those conditions. Over the years as, for example, leaf litter accumulates in deciduous forests, the biomass may increase because of the improved quality of the soil as a result of the previous growth. What we want to do on the Ecological Farm is to ensure that our ecosystem is an ecologically 'upgrading' one (Radcliffe, 1965): that its *net biomass* production increases year by year, not as the result of input from off the system, but as a result of its own previous growth and resources.

Much of the emphasis in conventional agriculture has been on increasing *gross* yield, at whatever cost—biological, environmental, economic, social, or in use of energy and scarce resources (see chapter 2). The aim of Ecological Agriculture is not to increase the yield at any cost but to increase the *net* yield per unit area. The use of modern knowledge of soil science and plant and animal breeding is clearly useful here, although its emphasis may be different from conventional agriculture. One way of assessing these might be in terms of energy budgets, as the biomass production of natural ecosystems is measured by ecologists. Alternative agriculture shows that there are some promising yields from similar systems. Boeringa (1980, p. 131), for example, says that yields can be as high as from a conventional system and figures I collected in Europe confirm this (see Table 5, p. 81). There is no doubt that the net yield and energy budgets are much healthier (see also pp. 78–80 and Table 6, p. 83).

4 It is Normally of Small Size

The reason for this is, firstly, because on the whole smaller unit size increases yield per unit area although labour inputs may be higher. This has been shown in Thailand and the Philippines (Schumacher, 1974 and Griffin, 1974). Secondly, small unit size generally increases the number of people owning or working land. This helps to overcome the social and political problems inherent in agribusiness agriculture (p. 47), provides employment and reduces the move to the cities. These considerations are particularly important in the developing countries, but such problems are also becoming apparent in the developed countries where, for example, only 1.5 per cent of the population of Britain owns land (Blaxter, 1976). This is a less equal distribution of land among the people of Britain than ever before. Land is one of a country's major resources, and its control by only 1.5 per cent of the population could be argued to be unjust and likely ultimately to lead to political instability.

The size of the farm will depend on the location, soil, topography, rainfall and so on. In most areas of Western Europe the maximum size might be about 50 hectares, but normally smaller if family and co-operative labour remain the basis of the working farm. In developing countries and Japan, the unit size might well be as low as 1–5 hectares; in marginal areas of the United States, and Africa, perhaps as large as 100 or even 500 hectares. What is happening in much of the world at the moment, in both the so-called 'developing' and 'developed' countries, is that the small peasant self-sustaining farmer is gradually being pushed off his land by the increasing price of land for cash crops, forestry, building and so on. All over the world there are increasing numbers of landless poor and sometimes starving peasants. In the 'developed' countries these people can join the welfare lines—hardly a solution but at least it keeps them alive. In countries without a welfare state system they are not so lucky.

5 The Farm Must be Economically Viable

Conventional agriculture, at least in the United States and Western Europe, is largely government subsidised, despite claims for its economic efficiency (pp. 33–6). Ecological Agriculture is intended to be economically viable without large government subsidies. This is in part possible because of a lower expenditure on inputs. There may, however, be a slightly greater expenditure on labour.

Economic viability is defined here as providing enough income to maintain all those working on the farm at a level similar to large sections of others in society, and to cover the farm's maintenance costs.

The aim is not to maximise profits at the expense of environmental ethical and aesthetic considerations. Cash crops, for example, would *not* be grown at the expense of the self-sustaining system, although this might make money. Animals would *not* be kept in intensive conditions and fed imported foodstuffs, even though this would increase income. Aesthetic considerations must be important in the design of farm buildings and the preservation of the countryside. The *conservation* of woodlands and hedges is therefore encouraged for their aesthetic value and also for sound ecological reasons (increase in diversity); and for economic reasons (woodland can be used in a self-sustaining way for farm maintenance and development and, if sufficient, to sell for cash). In addition various indigenous crops may be gathered for sale or consumption from the woods—elderflowers and berries, and blackberries, for drinks and jams, varieties of mushroom and toadstool for eating, herbs such as wild garlic for medicinal and culinary use (Mabey, 1972).

Ecological agriculture does not pretend to be the *panacea* for all economic problems. The aim is to increase net production and feed people and animals, look after the environment so that this will be possible indefinitely (provided the numbers of people and animals do not grow out of balance) and secondarily to generate profit, without all the other problems of modern agriculture.

Since conventional agriculture is directly and indirectly so heavily government subsidised, it is clear that economic problems are made for the farmer by government decision. Farmers farm the way they do because they are paid to do so by the government. Clearly, therefore, agriculture could be changed overnight by a change in the subsidy systems. Subsidies to encourage ecological agriculture would rapidly solve the majority of agriculture's problems.

It should be recognised that there is a difference in life-style between the urban salary earner and the rural farmer, and hence a difference in demands and consumption. Within an urban environment the worker must earn a large salary since all essentials (food, lighting, heating, rent, travel to work, property, leisure activities and so on) must be purchased. Because of demand they are usually more expensive than in a rural environment (take, for example, the cost of living in Tokyo or London and compare it with costs in rural/suburban Japan or rural Britain). If you are an urban dweller, all leisure activities except use of public parks must be bought. If these happen to be rurally based, such as fishing, shooting, enjoyment of the countryside, climbing, riding or hill-walking, they are relatively expensive. Even urban-based entertainment is costly. Hence the urban worker must have a reasonable cash income to live in an urban environment.

This is not necessarily true for the rural ecological farmer who, by contrast, finds that many of the essentials of living (food, often shelter, heating and so on) can be home-produced provided he is prepared to use locally available products. His leisure activities are normally country based and hence part of everyday life (fishing, walking, riding, climbing and so on). Hence his need for cash to sustain a reasonable standard of living is less than that of his urban colleague. The definition of 'malnutrition' or 'poverty' is often based on the *per capita* income (e.g. George, 1976). Clearly the rural farmer may in many cases be well fed and content without having reached the same income level as a city dweller, since his life-style, and perhaps his philosophy and value system, are often quite different.

Unfortunately most of the economists and sociologists who have considered some of the problems inherent in this dichotomy have failed to recognise it, because they themselves are totally urban based, and assume that everyone else would want to have these values too. It is not essential, indeed it would be positively harmful to both environment and often to the rural people, to 'provide all the facilities of the town in the countryside', as horrifyingly suggested by Johnston and Allaby (1977). Perhaps, however, it might be more to the point to try to provide at least a few more of the facilities of the countryside in the town, if urban dwellers are to realise the environmental difficulties of supporting their anthroexlusive life-style and belief systems.

It would be better to generate greater employment opportunities in the rural areas, firstly to provide products needed by the local community and secondly some income, even though salaries might not be as high as in the cities. This applies to both the developed and the developing countries, where rural employment in agriculture is declining as capitalisation of farms increases, sometimes with government subsidies (e.g. Blaxter, 1976; Dahlberg, 1979).

It is often assumed that food produced 'organically' or with 'higher welfare' criteria for the livestock is a 'better quality product'. It is also universally believed that it is more expensive to produce. Why should this be the case? The ecological farmer at least will not be spending on fertilisers, herbicides, fungicides, pesticides and machines to apply all these. She will not be spending on fodder and food for her animals. She may be paying more on labour, but why is it always assumed that this will be *more* expensive and that therefore she will have to charge more? Maybe she is just bad at it, or maybe she is just conning the market . . . a perfectly acceptable economic strategy. If she can get more for her products, why not play the economic game, if that is what she wants to do? There is little evidence that it is more expensive to produce goods in a self-sustaining way (see Table 2).

Whether or not the food is of superior quality is not the important issue here (although this has usually been used as the central argument for such food production by the 'organic movement'). One thing that is certain, however, is that where insecticides, pesticides and herbicides are not used, there is no risk of food contamination from them! The important issue, economically, is that the product will sell at the price the market can bear. If people are prepared to pay more for a small uneven tomato than for a large, evenly shaped one, then any economist would recognise that it would be acceptable business to grow small uneven tomatoes.

The 'superior nutritional value, superior price' organic growers have to my mind missed the much more serious reason for appropriate alternative agriculture: food for all without environmental dilapidation. After all, the main disease for most of us in the Western world is over-eating of foods too high in quality and the consequences of that, rather than starving or dying of poisoning. The quality testers for chemicals harmful to health may occasion-

Table 2. Comparison of the economic returns from beef suckler cows for: 1) an ecological farm (Little Ash) and 2) a conventional farm in the same area

COSTS FOR COW AND CALF

1) *Summer grazing* ecologically managed pasture: £60

Muck spreading, 1 ton/acre:	£5
Topping, harrowing, fence maintenance, etc.:	£10
	Total: £75

Winter feed: Fodder:

Cost silage making/big round bale (cutting, racking, baling, stacking, bagging	£3.45
Cost hay making/big bale cutting, turning, tedding, baling, stacking	£3.55
Eat 2 big bales silage and 3 of hay/winter = £3.45 × 2 + 3.55 × 3	
	Total: £17.55

Other food: Home produced grain: £60/acre rent, Mucking £5, ploughing £7/acre, cultivation and drilling £10 (seed saved from previous year), combining £20, storing £5 = £112/acre

Yield 1.50 ton/acre. Cost per ton £74.60

Cow given 25 kg	£1.80 say £2.00

Bedding:

Straw, Baling, storing £1.70/bale

3 bales/winter:	£5.10
Vet, deworming, etc.:	£12
Miscellaneous expenses, share of bull, etc. say	£20
Buying of second calf for double sucklers:	£120
	TOTAL COST COW AND CALVES: £251.65

INCOME:

If 1 calf sold @ 8 months = £350

Less Favoured Area subsidy for suckler cow: £74

PROFIT:

If 1 calf sold then = £172.40. 1 calf kept still

BUT Little Ash 2 calves/cow and both kept for 2nd year

Cost keep 2nd year of stores at same cost as cow without extra calf: £263.20

Sale finished 'Organic beef' with 10% premium at 20–28 months @ average £690. 2/cow — Total = £1380

Intervention payments finished beef/head @ £60 = £120

Total = £1500 for 2 years

TOTAL PROFIT/COW/YEAR: £750

Table 2. *(contd.)*

2) *Summer grazing*, conventionally managed. Rent: £60	
Fertiliser. 20:10:10. 7cwt/year/acre	£40
Topping harrowing, herbicides, etc.	£10
	Total: £110
Winter fodder	
Cost silage big bales bought £8/bale, 2 bales:	£16
Hay bought or made @ £10/bale. 3 bales	£30
	Total: £46
Other food:	
Bought in cakes and grains @ £7.50/25 kg; 50 kg	£15
Bedding big bale straw bought @ £5/bale 3 bales = £15	
Vet, deworming etc.	£12
Miscellaneous, bull, etc.	£20
	TOTAL COST COW AND CALF: £203
INCOME	
1 calf sold @ 8 months	£350
Less Favoured Area suckler cow subsidy	£74
	TOTAL: £424
PROFIT:	£424–£203 = £221
Cost keeping single calf second year same as cow: £203	
Sold finished at £590, Profit keeping to finished beef: £387	
Plus intervention payment around £60/head:	
	TOTAL PROFIT/COW/YEAR: £447

ally get it wrong, but we all eat food of a much higher nutritional quality than we actually need.

One thing is clear (whatever the relative price of products from different types of agriculture): a greater percentage of the price of goods from conventional high-input agriculture is in fossil fuels, scarce resources, highly capitalised machinery, transport, packing, fertiliser and other inputs, and less in labour. For self-sustaining ecological agricultural products, a higher proportion of the cost will be in labour and less in all these other costs, whatever their selling price (Boeringa, 1980; United States Department of Agriculture, 1980).

Having said this, we must remain slightly wary of the recent spiralling success of 'organic products' in Europe. Many farmers are now joining the 'organic' bandwagon in order to sell their products more expensively, without realising that there is much more to 'ecological agriculture' (if not 'organic') than not applying herbicides, pesticides and fertiliser. If it is to help solve the problems of agriculture in the future, these factors *must* be recognised or the same problems will simply remain.

Taking the conventional economic view of farming 'for money' will not change a thing, no matter what you grow and sell. The mind set has to change if it is to deal with all the problems, including the economic ones.

One example may serve to illustrate this vital point. I was recently joined by a young man who wanted to try and run the horticultural part of our farm. He had previously been running an 'organic' market garden. As we walked round the garden to discuss the things that needed doing, he said, 'Of course, you must realise that I shall be doing it all for money.' He found it very difficult to come to grips with the ideas of not buying or being given 'organic fertilisers' (seaweed, poultry manure), not using large amounts of imported paper, peat, wood shavings and plastic to suppress weeds, and doing the work with horses instead of tractors. Nine months on the garden is still suffering compaction from his enthusiastic use of a tractor when the weather was too wet. Our well-tried techniques, he maintained, would not 'maximise the income of the organic horticultural enterprise' . . . he could not afford to do it our way. However, he did not seriously examine whether this were really true, nor take into account the other environmental effects of his 'organic' horticultural enterprise. This indicated very clearly to me, yet again, the necessity for disseminating Ecological Agricultural ideas, particularly to *Organic* farmers and gardeners!

6 The Farm Should Process Most of the Products

One of the ideas of an ecological farm is to process the products on the farm and sell direct to the consumer. This will of course depend on what is grown and available on the farm and what are the farmer's and community's interests. One particularly important consequence of the development of local farm cottage industries is that the community can be supplied with many of the products it needs, and the products can be sold direct to the consumer, thus avoiding the middleman syndrome, so typical of 'industrialised countries', where a host of entrepreneurs make more out of each item than the producer, and the consumer pays for them all.

The lunacy of current development in this regard is well illustrated by the scallop industry in Mull in the Western Highlands of Scotland. Here local fishermen catch the scallops and sell them to a transporter. He takes them to the mainland in a truck (despite the fact that Mull is only a couple of miles off shore and they could be delivered by boat). The mainland factory separates the meat from the shell, sends the shells to be 'processed' (washed?) in Glasgow at another factory, and the meat to be 'packaged and frozen' at yet another in Fort William, about 100 kilometres to the north. Each of these factories then sends its own lorries to France (about 2,000 kilometres) where the meat and shells are sold to wholesalers and then restaurants. In the restaurant the chef puts the meat back in the shell and serves them . . . This is not an unusual story of the way 'business' operates. But is it environmentally sound, or sensible?

7 It Must be Both Aesthetically and Ethically Acceptable

Farming has had enormous effects aesthetically on the countryside since the 1939–45 war. The erection of cheap, large, ugly farm buildings, and in particular badly situated, ugly houses for farm workers and farmers, stand out like sore thumbs over much of Europe, Australia and the United States. The aesthetics and diversity of much of the landscape have also suffered from the removal of woodland, trees from pasture, and of hedgerows. This was brought to the public's attention by Shoard (1980) and others, who suggested that the responsibility for the landscape should be taken out of the hands of the farmers in Britain. It is remarkable that in a country where there have been severe planning controls for human buildings since the 1930s, agriculture has been almost exempt from such controls.

The countryside can retain a varied landscape, with wild areas, hedgerows and woodland which also yield their own harvest, contribute to the stability of the system, add beauty and give pleasure.

Farm buildings can be large, airy and easy to run provided the materials used to build them, their shapes and their siting and screening (or possibly disguising) are given sufficient attention. In addition, the farm workers' houses can be built out of local materials and properly sited.

Should one always build for decades or even centuries? Why is it always considered improper to build temporary buildings provided they are aesthetically pleasing—and why can they not be? Why should we burden future generations with our ideas in architecture? They may want to have some of their own! Perhaps we should examine the idea that all good buildings erected are necessarily long-lasting and 'properly built'. In other societies there are many examples of aesthetically very pleasing and useful temporary or movable buildings made from local materials—from benders and mud huts, to igloos and tepees. Let us open our eyes and not stick to the breeze block covered with pebble dash and, if you're lucky, a thatched roof with reeds imported from Hungary to get the real feel of Devon, or the 'Modern Scottish Awful' rural housing, when both areas have some superb traditional, ecologically sound building styles.

The landscape is essentially dynamic, and it is equally foolish to try and resist any change, which is the central philosophy often adopted by those apparently primarily concerned with aesthetic value, such as some National Parks. The rural environment *cannot* be 'preserved'—frozen in time as 'history'; by its nature it is constantly changing and evolving. The individuals or institutions who are concerned about this must aim towards helping the development of a self-sustaining dynamic, symbiotic living system in which developments are moulded thoughtfully into the countryside.

In this respect a cautionary lesson can be learnt from the hurricane of October 1987, which wreaked havoc on the South East of England's landscape by tearing up predominantly mature and old trees. This wealthy and densely populated area of the world had become very aware of the aesthetic value of trees in the rural and urban environment and had had 'preservation orders' placed on many old trees. As a result people had neglected to plant many young trees. In the hurricane many of the old but spectacular trees were blown down. The few younger trees were not. If a dynamic planting and culling programme had been carried out ten to fifteen years previously, there would still be woodland in areas that were flattened by the storm.

The ethical considerations are threefold and are discussed in detail in chapters 8 and 9. In the first place they largely concern animal husbandry on the farm. It is implicit in the system that animals be kept as far as possible within the type of environment, both physically and socially, in which they have evolved to live. In this way they are more likely to have more fun than misery, and optimise their net production (Kiley-Worthington, 1977). From the animals' point of view this is ethically more acceptable than intensive, highly capitalised, high-input animal husbandry.

Secondly, by keeping animals in ecologically and ethologically (behaviourally) sound environments, there is little or no need to import large amounts of protein substitutes from the developing countries, such as soya bean meal, fish meal and so on. Often there has been an incentive to grow these crops for export instead of for home consumption. It is in this way that developing countries have become less self-sufficient in food production in the last decade, so that more people (and their animals) starve. Actually, as we have seen, intensive husbandry provides food neither cheaply nor efficiently when looked at on a world scale (Dumont and Rosier, 1969, and chapter 2).

Thirdly, it is ethically unacceptable to destroy or jeopardise the environment and its future, not only because there are succeeding generations to be considered, but because animals, and possibly plants and the environment itself, may by their very existence have certain rights (Reagan, 1982; Naess, 1990).

With a self-sustaining approach to agriculture, it is possible for peasants and other farmers to maintain their independence of creditors and the agribusiness industry since they are not dependent on them for inputs (fertilisers, herbicides, pesticides, large equipment, and so on). The dependence is one of the main reasons for the lack of success of the Green Revolution in feeding the poor (see Dahlberg, 1979, and George, 1976, for detailed assessments). In Europe this is aptly illustrated by the work of Moeller who developed an alternative agriculture in Switzerland for this very reason, and whose ideas have been largely responsible for the continued independence of the small farmer, his success and therefore his voting power in Switzerland today, despite encouragement by the conventional agricultural services to increase inputs.

Such an approach has been criticised in different ways. It is said by some that it represents too closely the rural grind of the eighteenth century. Others maintain that too many people could be attracted back to the countryside, and that in any case it cannot work. All these criticisms are unfounded. In the first place, rural living demands a different life-style from city living, with different economic and other requirements. Others (Ulbricht in Boeringa, 1980, and Hill, 1981) have stressed the importance of the holistic approach. The awareness of men and women of themselves, their environment and their relationships with other living forms and the entire biosphere are implicit in Ecological Agriculture.

Yet others have been struck by what they call the 'pseudoscientific' justification for certain practices put forward when the ideas are essentially non-scientific. Science does not have to confine itself to non-holistic investigation and thinking; it is after all the pursuit of 'systematic and formulated knowledge' (*Oxford English Dictionary*). Surely a good scientist will draw in knowledge and information from any relevant source to try and understand her problem; she must essentially be interested in information from every discipline, although she may have to make judgements about the degree to which each statement is true or false, using empirical tests, perhaps, or other rational arguments and aspects of the 'scientific method'.

I am concerned here principally with pointing out a 'systematic and formulated approach' to the defining and operation of Ecological Agriculture. There is no conflict between a rational scientific approach and a holistic/ecological one.

Many people who would agree that the outline given of Ecological Agriculture sounds acceptable, would add that it is impossible. But there are some 30,000 farmers in Europe farming this way to some degree, and many millions in Africa, Asia and South America. Their farming systems could, in many cases, be greatly improved, and hence their nutritive level and possibly incomes, but in most cases *they need not be radically changed*. As we have already discussed in chapter 1, it is in the Third World that such an agriculture has great possibilities. If we encourage the multinationals' expansion into the Third World, and with it the peasants' increasing dependence on expensive inputs and hence creditors, and the growing of cash crops to sell to the West (at the cost of growing their own food), the Third World will then have very little chance of feeding itself in the future. The redistribution of food from the North to the South has yet to overcome massive problems of politics, inefficiencies and corruption and, despite increasing donations, it is unlikely ever to become fully effective (Marstrand and Pavitt, 1974; Erlich *et al.*, 1977).

One of the points of Ecological Agriculture is that, like ecology itself, it conforms to certain basic principles, but over and above this there is room for enormous variation. Ecological Agriculture must therefore be adaptable to different environments, climates and geology, and to different social, political, aesthetic and even, perhaps, ethical constraints and approaches. It may be that nomadic pastoralism is the most appropriate ecological agriculture in certain arid areas, but self-sustainability for the people and their animals over the nomadic area must be retained if it is to be truly appropriate. By contrast, there might be a case for managing certain areas, such as the prairies, in large unit size with relatively high capitalisation, but again self-sustainability would be the goal, ideally within the farm, although this might be extended in certain cases to the locality, shire or country.

4 Does Ecological Agriculture Work?

In 1976 I was fortunate in being awarded a Winston Churchill Memorial Travelling Fellowship to enable me to travel around Europe and look at farms that were considered Alternative. These were organic and in particular biodynamic systems. Wherever I went in France, Holland, Denmark, Germany and Switzerland, I tried to find farms that fulfilled at least a few of the criteria outlined in chapter 3. During this survey, the need for a well-defined agriculture which tackled all the problems described in chapters 1 and 2 was clearly emphasised to me, despite the language difficulties. At the same time it became clear that even on these farms, none of which could seriously be considered truly ecological although all of them confronted some of the problems, there were some quite dramatic successes. It became even clearer that conventional agriculture and the arguments constantly made against these types of agricultural alternatives were often false. I was able to collect many ideas, information and figures on how they worked to help in the formulation of Ecological Agriculture. Combining this with our own experience and some further figures highlights the most important considerations and reveals whether or not the ideas of Ecological Farming are likely to be practical.

Arable Management

One of the priorities for the correct running of an ecological farm is to assess the ratio of land for arable to that for livestock. The latter, apart from being productive themselves, must provide sufficient manure to maintain and improve the land for arable crops.

The average percentage of arable acreage for the self-sustaining farms that I visited (by this I mean farms where little or no inputs were bought in) was about 30 per cent (Table 3). This would naturally vary depending on a large number of environmental factors, including the farmer's own interests (does he prefer to concen-

Table 3. The percentage of arable area to pasture on self-sustaining 'organic' farms in Europe

Farms (13)		*Total*	*% Arable*	*Grass*	*Other enterprises*
Renard	France	32	25	75	Dairy and cereals
Lotte	France	70	18	82	Dairy and cereals
Monsies	France	120	15	85	Dairy and cereals/ etc.
Maynard	France	115	13	87	Dairy and cereals
Vortmann	Switzerland	50	56	54	Dairy, fruit, cereals, veg.
Fry	Switzerland	30	53	47	Dairy, fruit, veg.
Blazier	Switzerland	16	37	63	Dairy, fruit, veg., wheat
Schendecker	Switzerland	18	25	75	Dairy, fruit, veg., wheat
Vogel	Switzerland	14	22	78	Dairy, fruit, veg., wheat
Texel	Holland	88	17	83	Dairy, sheep, cereals
Guepin	Holland	115	60	40	Dairy, veg., wheat
Moeller (Denmark)		55	36	64	Dairy, cereals
Scharmer (Germany)		98	29	71	Dairy, cereals, veg.
Mean			31.2%	68%	

trate on livestock or arable?) but this percentage figure gives us a guide. High-input farming can increase this proportion even up to 100 per cent and is a common development on the prairies and in parts of East Anglia in England. In a mixed farm (arables, grass and animals), to retain the self-sustaining character of the enterprise one should reckon on around one-third of the land in arable crops. Self-sustaining systems with a much higher proportion of arable have been developed in China by the careful recycling of all waste products including human sewage (Blobaum, 1975). This approach could increase the percentage arable on ecological farms in Europe . . . if it is desirable.

1 *Rotations*

In order to maintain the self-sustaining system, nutrients and trace elements must be recycled and nitrogen levels kept high, or increased. This can be done using rotations, as was discovered in the Middle Ages in Europe. However, the use of rotations to maximise yields within the self-sustaining system has received little attention *recently* from conventional agriculture, although in many developing countries in particular, rotations are still widely practised, particularly in subsistence systems.

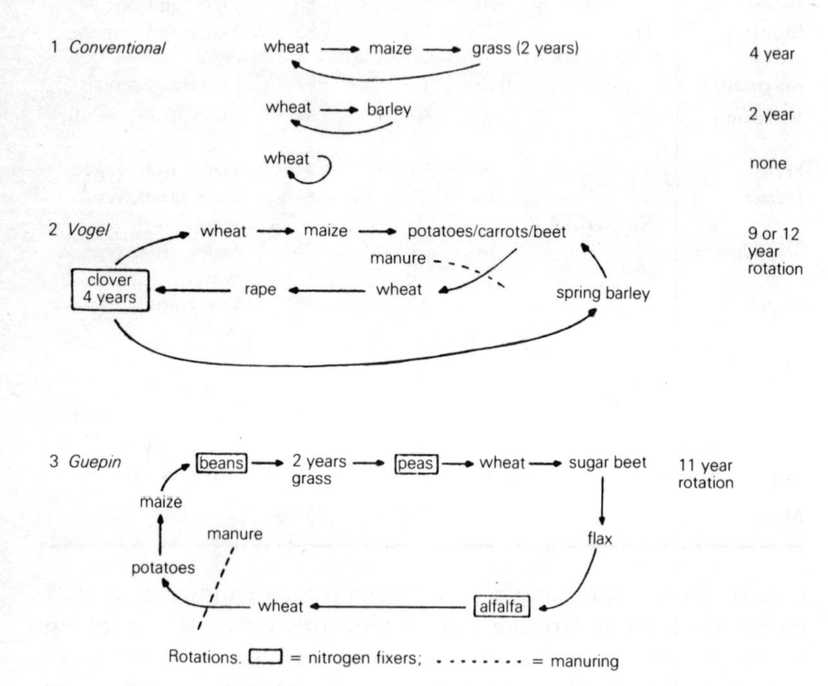

Figure 5. Rotations. Conventional rotations (1); and more complex rotations on a mixed organic farm in Switzerland (2); and a mixed organic farm (dairy and arable) in Holland (3).

On the whole, the organic farms I visited in Europe during the course of this work had complex rotations, of seven to twelve years (Figure 5). The organisation of these revolved around having several crops to increase or maintain available nitrates, such as the Leguminosae (which fix nitrogen from the air), and also green manuring crops that were ploughed in, such as rape. The farmyard manure (FYM) had to be spread on the arable land before the appropriate crops, since some crops such as roots do not respond well to FYM, whereas potatoes do. FYM was not applied every year, as the illustrated rotations indicate. One old technique I saw practised was to grow a mixture of cereals, beans and peas—'dredge corn'. These are harvested together, ground and fed as a balanced, reasonably high protein ration to dairy cattle in winter.

The rotations on ecological farms will inevitably be more complex and require much more planning, and often, perhaps, more thought, than on a conventional system where lack of these requisites can often be compensated for by bought-in fertilisers, herbicides and pesticides.

2 *Manuring*

The treatment of manure is vital in an ecological system, in order to ensure that the maximum levels of nutrients are being returned to the soil. There are several different techniques which are practised. The first approach is that used by the biodynamic farmers. They compost the manure, turn it, and then treat it with a special preparation to speed up breakdown. Results are good from such a system although the reasons are not fully understood. Another approach is to separate solid from liquid manure and pump the liquid straight out onto the grass. In this way an early 'bite' can be obtained in the spring for dairy cattle, because this liquid manure acts almost as quickly as inorganic fertiliser in encouraging growth of grass. Another approach was developed by the Moellers in Switzerland. This was to apply fresh manure as a top dressing, and it also works well.

The rate at which manure is applied varies enormously from farm to farm. The Cadiou *et al.* report (1975) from French farms quotes the figure of 12 tonnes per hectare per annum for arable

Table 4. The amount of farmyard manure applied per hectare per annum by organic farms in 1) France, and 2) France, Holland, Denmark and Switzerland

	Polyculture	*Market gardens*
1. C. Le Pape	12 T	75 T
2. This study	4–6 T	50 T
Variation	1.5–15 T	

crops and 75 tonnes for market gardens. I found that on the whole rates were much lower, but varied (Table 4). All these results show much lower figures than for conventional farming, where manure or slurry may be disposed of onto the land at the rate of 100–200 tonnes per hectare. It is possible that such high doses might have a degrading effect on the soil and therefore the yield (Rauhe and Knappe, 1971).

One of the main variables is how much manure is available. This in turn depends on the ratio of arable to animal land (see Table 3), and whether or not animals are stabled and bedded in the winter. We find that 2.5 tonnes per hectare per year are sufficient to retain a carrying capacity of 4.35 bovine units per hectare and a cereal crop every five years, but much more research would be very useful here.

3 *Weed and pest control*

Specialised monoculture systems are usually relatively unstable. Annual cultivation of the same crop in the same location leads to the build-up of particular weeds and pests, and hence pesticides and herbicides are used in increasing quantities to control them. Attacks by pests, and weed problems in arable crops grown on diversified low-input systems, were found to be minimal, although no chemical treatment was used.

Plant breeders have selected for higher yields, often by inbreeding—a quick way to select for specific characters. This reduces genetic variation and often results in reduced viability and ability to withstand competition and pest attack (e.g. Maynard-Smith, 1966; Jewell and Alderson, 1977). This is well illustrated by the High Yielding Varieties (HYVs) of rice and wheat (e.g. Dahlberg, 1979) which are dependent on chemical inputs to survive. Biological and integrated pest control methods are needed for diversified self-sustaining farms (e.g. Smith and Reynolds, 1966; Woods, 1974; Way, 1977) and needed urgently in developing countries with little access to chemicals (Haskell, 1977).

Diversification, which often increases stability, plus the use of seed selected on survival criteria, helps to overcome these problems without having to resort to the pesticides and herbicides which may have unforeseen effects on the ecology, and thus on farm production (Carson, 1962; Dasmann *et al.*, 1973; and chapter 2).

There are other techniques which help to reduce the problems of weeds and pests in arable crops. These are:

1 interplanting,
2 close planting,
3 companion planting (the planting of sympathetic species together),
4 the development of mechanical weed controllers.

In Switzerland and Germany, for example, harrows with adjustable tines have been developed, which allow harrowing of a cereal crop. This is very successful in destroying annual weeds. The harrows can be used until the ears of the cereal appear. Another ingenious European invention is the development of a flame-thrower carried behind the tractor, which burns off the entire cereal crop at the three-leaf stage, after which the crops regrow, but the annual weeds do not.

Pest control could be much further developed using biological controls such as predators and parasites—for example, ladybirds and lacewings to control aphid attack, and beetles against weeds (Huffaker, 1958). Here we must be very cautious, as inadequate study and understanding of the pest's ecology can lead, at best, to masses of misspent public money and, at worst, to disasters (e.g. Brown, 1961). An integrated pest control, involving 'commonsense cultural practises', biological control and, possibly, short-term biodegradable pesticides, could be the first step. Decoy plants could be used to attract pests away (e.g. Stern *et al.*, 1969) as well as parasites—for example, calcid wasps have been used successfully to control a number of major pests (Odum, 1971).

Despite these measures the crops will be subject to pest attack from time to time. The peaches, pears and cherries produced are not necessarily identical in size, shape or taste, unlike those of conventional agribusiness. Yet these varied fruit sell at a premium for several reasons, not least because they are free from chemical tainting and poisons; something the consumer is becoming increasingly aware of. Slight attacks of rust or smuts, for example, in organically grown cereals are not unusual, but the attacks tend not to spread rapidly through the whole crop as they do in a high-input, specialised monoculture (e.g. discussion in Dahlberg, 1979, pp. 81–3). Organic farmers say that probably every pest can be found on their farms, but very rarely, if ever, reach epidemic proportions, in contrast to what happens in conventional agriculture based on monocultures, where more frequent sprayings using constantly changing formulae are needed to combat epidemic pest attack.

Resistance to disease may be enhanced by selection of home-produced seeds, since these are adapted to local conditions. In this case a varied genotype (characters in the genes which may not be displayed in every generation) is being selected for, on the basis of survival criteria rather than just high yields.

There are other advantages to not using herbicides and pesticides. For example, vegetables and vines can be grown closer together since it may not be necessary to get a tractor and sprayer between the rows, the weeding being done with horses instead. Thus yield can be increased. In addition, the reduction of microbial activity, environmental pollution effects and the accumulation of poisons along food chains are avoided.

Interplanting of alternate rows of different crops, or just the broadcasting of mixed seeds, such as the planting of maize, beans and sweet potatoes practised in much of East Africa, reduces the need for weeding and pest control, provided the different species are planted at the appropriate time of the year. Catch cropping (planting a different crop between the major crop, thus increasing the production) as practised successfully in the Philippines (see Griffin, 1974, for discussion), as well as controlling weeds and pests, increases the yield per acre and the variety of the diet. More experimental work needs to be done on these kinds of approach.

4 *Yields*

Once the self-sustaining system is developed (which can take several years after changing from a system using high inputs), the arable yields are only slightly lower than the conventional high-input yields. Since there is a different crop each year, the yields will vary annually. Until we have yields presented as *net energy production* for each system, yield figures are not totally comparable (Table 5).

Because energy-consuming materials such as manufactured fertilisers, herbicides and pesticides are not used on this system and the products are sold direct to the consumer without elaborate processing in factories, the energy budgets are obviously more balanced. Manual labour energy input must also be calculated and is likely to be higher in the organic systems. In chapters 5, 6 and

Table 5. Gross yields (in tonnes/hectare) compared for conventional and organic high-input farms in Europe

	Wheat yields	*Oats*	*Barley*	*Beans/ peas*	*Sugar beet*	*Potatoes*
Average:	4.2	4.4	4.2	3.0	50	26.5
Range:	3–5.7 (6–7.8)	3–6 (4–7)	1–6 (4–7)	2–4 (3–5)	–	25.30
Number:	8	11	7	4	2	2

Brackets = conventional farm yield in similar localities.

7 we have made some steps in trying to assess energy budgets off our various farms (Tables 10 (p. 106), 16 (p. 140), 17 (p. 158).

Others have found the yields of organic systems to be as high as in high-input systems (e.g. Mattingly, 1974; Cadiou *et al.*, 1975; Boeringa, 1980; United States Department of Agriculture, 1980). It must be remembered that the miracle 'HYVs' may not be appropriate for this system as they may have physical, cultural, social and economic problems that come in their train (e.g. Dahlberg, 1979); what we want are hardy, adaptable, reasonably high-yielding varieties.

Although yields of large units in the Western world have increased considerably (with higher inputs), there is evidence that the energy budgets are better and that the biomass (total growth) per unit area increases with diversification and the smallness of the holding—for example, in Thailand and Sri Lanka, Indonesia and Africa (Griffin, 1974; Horath, 1977). The back garden is often the most productive area; it may be, particularly with all the other uses that people and other animals wish to make of the land, that the more net productive, small-scale unit will be the most suitable for much human food production.

On the whole it is fair to conclude that the arable side of ecological farming is successful and shows great promise for future development, in both the developed and underdeveloped worlds, despite the notable lack of government backing and research effort so far.

Grassland Management

The management of grass on the organic and biodynamic farms has often not advanced as much as their management of the arable and horticulture. Grassland is treated more as a cereal break than a crop in its own right and there is often little management of it in any consistent way. Information on grassland management from other surveys of organic farms is also sparse (e.g. Cadiou *et al.*, 1975; Boeringa, 1980; IFOAM, 1986).

The aim of Ecological Agriculture's management of grassland is to set up a self-sustaining system. Clearly, animals sold off the system represent a net loss of nutrients which must, in some way, be replaced. We have already mentioned the return of all animal and human wastes to the system. The resulting high humus levels together with the legumes in the herb layer ensure that nitrogen levels can be maintained at a level sufficient to encourage good growth (Table 6).

Since animals' bodies, like those of humans, are composed largely of water, the most important loss from the system when animals are sold off every year is likely to be phosphates in the animal bones. However, we found no change in our phosphate levels over our ten years at Milton Court (see p. 104). Nevertheless, to ensure that there is not a gradual decline over maybe decades and centuries, it is sensible to ensure that the bones are returned either as bone meal or decomposed. Alternatively, the animals sold for meat can be slaughtered on the farm or in the local abattoir and the meat sold off the bone.

Loss of trace elements from the grassland might be a problem, but by having a diverse sward including deep-rooting plants and plants that concentrate particular elements, this can be avoided.

Evidence for the differential concentration of elements in different species and even different parts of plants is accumulating. The Soil Association maintains that camomile concentrates calcium and potassium; dandelions, calcium, copper and iron; nettles, phosphate; plantain, potassium and copper, and so on. Research done on Milton Court farm (Holder, 1978, and Table 7) found considerable differences in the concentration of calcium, potassium and nitrogen in several species and parts of species in the grassland.

Details of the exact percentages of such plants in the herb layer, to maximise recycling of trace elements and provide maximum

Table 6. Carrying capacity of grassland from conventional high-input farms (Con.), mixed organic farms in Europe (Org.), multi-species ecological farms: Milton Court (MC), Druimghigha, (Dr) and Little Ash Eco-Farm (LA)

Figures in 'bovine units', i.e.: calculated equivalences to one adult cow and calf

Ley cereal breaks					*Permanent pasture*				
I species grazed		Eco-agriculture			1 sp.		eco-agriculture		
Con.	Org.	MC	Dr	LA	Con.	Org.	MC	Dr	LA
Additions: 250–500k t/hec NPK	1–3t rock Ph, seaweed extract etc.	FYM 2t/hec			same				
Number farms: 10	9	1	1	1	8	9	1	1	1
** Summer food+10–60% winter	winter	Including 80–100%			Summer +10–60%		Inc. winter fodder: 80–100%		
Bovine units: (average) 2.25 range: 2–3	1.46 1–2.5	2.4	2	2.6	2.8	1	2.4	1.8	2.8

NOTE: not only is the summer carrying capacity of the grassland as high as when inorganic fertilisers are used, but the ecoagric-multi-species grassland management supports the animals to a greater extent through the winter than either 'organic' or conventional management.

graze and palatability for the grazing species, remains to be worked out. However, these results are promising.

Research on the African savannah has shown that in some cases a greater biomass is carried by having a variety of grazing and browsing species to use the different niches in the diverse habitat (Dasmann *et al.*, 1973). Because the different grazing species have different eating habits, and select different types, parts and stages of growth of plants, they are less competitive for the grazing resource and are able to use all the food resources in the area, which would be less possible if the area was grazed by only one species (Johnstone-Wallace, 1937; Bell, 1970).

A similar type of approach might be applicable to the manage-

Table 7. Nutrient levels from various dicotyledons from Milton Court Farm (Holder, 1978)

Plant		*Potassium*[a]	*Calcium*[a]	*Nitrogen*[a]
Sorrel	Leaf	4.467	0.321	
	Stem	6.176	0.353	1.335
	Root	0.473	1.067	
Stinging nettle	Leaf	2.255	3.826	
	Stem	2.546	0.851	3.026
	Root	0.897	0.325	
Sheep's parsley	Leaf	3.287	0.954	
	Stem	3.026	0.764	1.487
	Root	1.247	0.368	
Daisy		1.225	0.965	3.795
Dandelion		1.718	0.699	1.393
Clover		2.156	0.937	4.732
Spear thistle	Leaf	2.816	1.707	3.709
	Stem/root	2.894	0.590	
Usual limits:		0.5–3.0	0.3–2.5	1–3

[a]% dry weight.

ment of agricultural grazing systems, whether they are extensive, such as the prairies, mountains and moorlands, savannahs or steppes, or more intensive, such as lowland agricultural land of permanent pastures or leys rotated with arable crops. There is substantial evidence that multi-species grazing, normally only extended to beef and sheep, increases the carrying capacity (e.g. Peart, 1962; Clarke, 1963; Culpin *et al.*, 1964; Hamilton and Bath, 1970). It is time this information was used and extended by including other species in the grazing system—cattle, sheep, horses, llamas, geese, to name a few.

Of particular importance in regard to multi-species grazing are the differences in defecation patterns of different species, and the avoidance of faeces, particularly of their own species, and hence the reduction in the frequency and quantity of infective parasites as compared to a one-species grazing. There is also evidence to suggest that many other factors are important in this respect, such as differences in species or parts of the plant selected by different age groups (Leaver, 1970) and possibly even sexes. Social facilitation (do-what-others-do) affects grazing behaviour (e.g. Hardison *et al.*, 1954), as does herding and driving.

The suggestion is that net production might be greater if:

1 the systems were designed to be as self-sustaining as possible and therefore composed of many floral species,
2 several different animal species were grazed.

In order to maximise production along these lines we have developed the following management strategy:

In the first place the grassland is divided into small paddocks separated by hedges and shaws. The paddocks are then grazed when the grass is approximately six inches long in rotation, in two or three phases:

1 Lactating cows and other animals requiring long grass, and the pickings for growth, milk production or reproduction.
2 Animals that will eat the sward when it is a little shorter, and require food to maintain their bodies plus a little extra (e.g. fattening cattle, mares and foals, working horses and breeding stallions, flushing ewes or pre-parturant or early lactating ewes).
3 Animals requiring maintenance, and/or able to graze short grass (e.g. ponies, some horses, sheep and llamas).
4 Animals able to graze very short grass and maintain or increase body weight (e.g. some sheep and llamas).

For more than 40 years it has been known that grazing the grass very short encourages its growth (e.g. Proctor *et al.*, 1950). After the animals have been taken off or, in some cases, while some are still on the pasture, it is topped: the seeding heads of various plants are cut off to reduce the seeding and hence control the spread of some dicotyledons or 'weed species' such as thistles and docks. It is then harrowed to spread the faeces and shut up as the animals rotate to the next paddock. The previously grazed paddock is then grown up and cut for hay before being grazed again. This hay break provides winter feed and allows time for the breakdown of the faeces which might otherwise affect selection by the animals in the next graze. It also encourages a heterogeneous (species-diverse) sward by allowing species to flower and seed. In this way, apart from its conservation value, an evenly grazed very species-diverse sward is developed, including many wild flowers. Farmyard manure is spread between one grazing period and the next, either annually or bi-annually, at the rate of 2.5 ton/hectare/annum.

Clovers and other nitrogen-fixers such as vetches are encouraged. In order to ensure the right amount of nitrogen, we like to have clovers in at least 30 per cent of the sward in summer, the dicotyledons in another five per cent, and various species of grass selected on grounds of palatability, appropriateness for the area, winter greenness (to increase the length of the grazing season), hardiness and production. On our farm, the carrying capacity and production have improved considerably (see Table 6, p. 83). These carrying capacities are compared with other local farmers' grassland carrying capacity managed for one species and applying N, P and K at rates varying from 50–100 kg/hectare/annum (see chapter 5).

The grassland on our first experimental farm was long-term leys which were converted to permanent pasture. By re-seeding with appropriate seed mixes and adding geese to the system (which is not done at present), it would be quite feasible to improve the carrying capacity by two more bovine units per hectare. The extra cost of production with conventional N.P.K. (nitrogen, phosphate and potassium) treatment of the grass, and with monoculture grazing, would be approximately £25 per hectare. The gross production for the whole farming system of the ecological farm, of which 27 acres were grass managed in this way, was £700 per hectare for 1978–9.

This grazing system has also been adapted to a marginal area—'a wet desert' on a Scottish Hebridean island. It has proved to be an essential ingredient in beginning the biological upgrading of such an area (see chapter 6).

Animal Husbandry on the Ecological Farm

Inherent in the notion of ecological farming is the use and even creation of many plant and animal 'niches' to increase diversity. Thus animals are selected as appropriate to fill available niches. Not only suitable species, but suitable races or types of animals must be selected for a particular local habitat.

Such an idea is not restricted to physical habitat, but includes the types of food fed, and the psychological requirements of the specific species. For example, the degree to which the species is social, the size of the group and its age and sex composition in which it normally associates when given the option, must be known and understood, and so far as possible provided for. Information on the physical requirements we can gather to some extent from the animal's external and internal morphology (body shape) and physiology. Behavioural information must be gleaned from the study of feral or free-ranging groups of our domestic animals and their close relatives. Under such a system, then, we could suggest that man becomes again a symbiotic partner with his farm livestock, rather than parasitic on them. This is a complex subject and is therefore discussed further in chapters 8 and 9.

Herbivores such as cattle, sheep and horses, which have the ability to convert cellulose to protein, are fed diets high in fibre and roughage which are produced on the farm (grass, hay, silage and straw) and not fed high levels of imported high-protein foods in order to increase growth rate or milk yield. In this way milk, meat and wool can be grown off grass (which is a necessity for the rotations).

The emphasis on high-concentrate feeding, even of beef cattle, is illustrated by the one million acres of pasture which, over the course of the last two decades, have fallen into disuse in the United States, while more and more beef is raised annually in feedlots on imported concentrate feeds (Dumont and Rosier, 1969), and more and more of the Amazon forest bites the dust for the quick, short-term monetary gain of selling hamburger beef to the United States. Until 1986, the EC policies were geared, apparently largely under the influence of British agricultural lobbies, to increase output by increased feeding of concentrates.

Designing and keeping animals in environments suited, in both the physical and psychological senses, to what they have evolved to live in has the added advantage that it reduces physiological stress. There are other advantages, such as keeping animals in appropriate 'niches', which are discussed in chapters 8 and 9.

Animals such as pigs and hens are kept as scavenging omnivores, utilising as food various wastes of field, house and farm such as slops, whey, buttermilk and weeds, and are kept in small numbers, either free range, or more extensively managed than in modern pig and poultry husbandry. No high-protein foods are purchased for such animals, only food produced by the farm. Production may be lower, but the net cost is very much less, both on a world and local economic scale. Risk of disease transfer, because of which feeding of untreated slops to pigs in the United Kingdom has recently been outlawed, can be overcome if necessary by cooking the wastes before feeding, as was often traditionally done. This is not an enormous task if numbers are small.

Disease transfer when all food is produced on the farm, and the animals are kept in non-intensive husbandry conditions and relatively small groups, is considerably less than under modern conventional high-input intensive husbandry conditions—for example, there is less mastitis (Ekesbo, 1978) and no lameness.

The reason why these restrictions on the use of slops or waste food stuffs have been introduced in Britain is because of the increasing risk of swine vesicular disease as a result of intensification of pig husbandry. Most pig herds in intensive husbandry systems are treated with various prophylactics including, in some countries, antibiotics in foodstuffs. These appear necessary to maintain the system. By contrast, most 'backyard' pigs are not treated with either growth promoters or prophylactic medicines to the same degree, and there is no evidence that they suffer more disease.

Ducks and geese can be kept on ponds and wet areas, goats on very rough pasture, and unconventional species such as red, fallow and roe deer can be kept and cropped off moorland and/or forest as appropriate.

Appropriate Animal Breeding

I have argued for diversification of animals on the farm unit in terms of species, but also in terms of age groups. Arguments can be made for having both male and female animals on the farm—for example, the presence of a bull increases conception in cows even when AI is used (Kiley-Worthington, 1977, pp. 44–8). For this type of agriculture the animals, like the plants, will be selected on different criteria from those used for conventional high-input farming. The chief of these criteria will be:

1 The animals must be able to survive and grow on low-concentrate diets, and be able to utilise available products adequately, rather than being selected for high performance on high-protein diets. The modern Friesians, for example, are selected predominantly on milk yield as a result of high-concentrate feeding. Such animals perform very badly and are particularly prone to disease, foot problems and infertility when fed on low-concentrate diets, as I know from personal experience. Thus many of the modern breeding programmes may

well be inappropriate for the future and we may need to return to breeding from the older indigenous minority breeds. The breeders themselves are now realising this and are having to select on the grounds of five to eight criteria—not just milk yield. Back to natural selection (British Cattle Breeder's Club Digest, 1974).

2 The animals must be disease-resistant, and require low levels of medicaments. The increasing occurrences of diseases such as sub-clinical levels of mastitis and lameness are disturbing and are the results of bad environmental design, selection on the basis of high yields only and emphasis on higher production by higher concentrate feeding.

3 The animals must be adapted to the particular environment. Local varieties are likely to be more successful under this type of management system than imported ones, since they have evolved in that environment. For example, in Sussex, Sussex cattle would be most appropriate, South Devons in Devon. There are some surprises here, however, for we took our South Devons to Mull. The first year we all had a bad time adapting to the much harsher conditions, but by year six the South Devons were out-producing and had a higher fertility than the indigenous Galloway and Galloway/Shorthorn crosses we had bought when we arrived (Table 8).

4 In some cases it will be useful to have multi-purpose breeds—for example, dual-purpose cattle producing both high-quality milk and beef; horses for riding, driving and working; poultry for both meat and eggs; sheep for wool, meat, and milking, and so on. In these cases, therefore, selection will be away from a tendency to specialise.

5 Animals must be behaviourally suited for such a system. For example, all poultry will be expected to sit on their own eggs. A high egg-producing hen which retains the ability to go broody will therefore be required.

Turkeys will be expected to copulate normally and produce fertile eggs rather than relying on AI, as is the case in the common white commercial turkey.

Cows will usually suckle and look after their calves, and must therefore have appropriate maternal behaviour and produce sufficient milk. The latter is often not the case with the pedigree Hereford, for instance, where a surrogate mother may

Table 8. The economics and production compared for Galloway cross cattle typical of the area and South Devon cattle taken to Druimghigha.

Both groups were out all year with shelter they could visit in the winter and where they were fed silage, straw and hay. None were fed any grain or concentrate foods unless they were ill.

Galloway cross	*South Devons*
beef shorthorn and blue greys 15	14
All spring-calving cows all run with bull with free access to hills in summer, and on 100-acre hill pastures in winter	
Suckled and raised own calf	Suckled and raised own calf and adopted bought-in Angus Friesian/Hereford calf
Store calves sold in November to December Price: £200–£250	
Money produced per cow:	
	£200–£250 × 2 = £400–£500 – bought calf: £70
£200–£250	= *£330–£430*

have to be given to the highly bred pedigree calf because his own mother does not produce enough milk even to feed him well.

These are often the cows that sell at the highest prices to 'improve' herds in far-flung areas of the world such as South America, South Africa, the United States and Australia.

Sows must have a chance of being able to search for food by rooting, and to nest-build and farrow without rolling on their young, rather than being restrained in pig-farrowing crates.

Natural behavioural strategies should be allowed. For example, normal calves have well-developed shelter-seeking behaviour; they are therefore able to survive and do well in difficult weather conditions rather than being brought in to confined and often dirty and crowded barns.

Today the modern high-input, intensive husbandry units often encourage or require the opposite behavioural criteria, which are therefore being selected. In such units it is probably important for the animals to be able to 'switch out' (do nothing), since confinement, crowding and boredom may cause 'distress' (Kiley-Worthington, 1977, and chapters 8 and 9).

The more complex and changing outdoor environment requires a re-establishment of, for example, appropriate grazing strategies which encourage correct nutrient selection. This may involve rapid learning which would be selected for.

Many of these behavioural characteristics have not as yet been totally lost from the gene pool, but they soon may be unless appropriate action is taken.

6 The aim will be towards high quality of animal products, but also a greater variety. Milk with a high butter fat, for example, would be more appropriate where milk is being processed on the farm. This is because some cream can be taken off and eaten directly or turned to butter, while the skimmed milk will still have over 3.0 per cent butterfat, acceptable for household use.

A degree of hygiene is necessary and is controlled in Europe with the licensing systems. At present, in Britain at least, the authorities are not in favour of small producers. They have insisted on pasteurisation of milk before sale, which, because of the cost of the plant, has put many small producers out of production. Nevertheless, even under such a system it is still possible to obtain a licence for small-scale milk products.

Beef, pork, poultry and sheep meat can be sold directly to the public, which results in lower prices for the consumer and a greater percentage of profit for the farmer. There exist in Europe necessary laws concerning the humane slaughter and inspection of meat for small-scale producers. Mobile abattoirs with professional operators would be an improvement in the future, reducing the fear and trauma of the animals travelling and being slaughtered in a strange place.

7 The selection for breeding animals with longer working lives is important for the small producer, because although his profit per animal may be greater as a result of lower inputs, possibly fewer problems of disease, lower capital expenditure on buildings and so on and direct marketing to the public, nevertheless

he will have fewer animals to sell. However, keeping animals in environments closer to those they have evolved to live in, and with low levels of physical and psychological stress, will help to increase length of life (see Table 1, p. 39).

8 Special requirements for the crafts associated with the farm will be fulfilled by keeping particular varieties of animals, for example ewes for particular fleeces (Jacobs, Wensleydales or hairy coated Wiltshires), angora wool goats or rabbits, or llamas, guanacos and alpacas for spinning wool.

 Feathers of different colours, or special types of hides, might also be required and could be produced.

9 Finally, yield will be selected for, given that the animal fulfils these other criteria.

 The farm must be economic as we have already mentioned, but the prime reason for its existence will not be to make money, but rather to maximise production to feed people and animals, at as little energy, environmental and social 'cost' as possible.

For selection along these lines, the major breeds today are often inappropriate, and one must look farther afield to find local or 'old-fashioned' breeds and begin to select from them. The danger is that if the present monopoly in agricultural thinking, and animal production in particular, continues, we shall rapidly lose the genotypes of some of the animals which will be more appropriate to this type of selection and farming (e.g. see Jewell and Alderson, 1977). The percentage of the British national herd that were Friesians rose from 64.2 per cent in 1965 to 76.3 per cent in 1970, and is still on the rise, while the percentage of crosses and breeds other than the dominant two dropped from 3.8 per cent in 1965 to 2.5 per cent in 1970 (Craven and Kilkenny, 1976).

Ethics and Farm Animal Husbandry

It is argued that since intensively raised animals often grow well, put on weight fast and are provided with shelter, warmth and plenty of food, they have all their requirements catered for, and that therefore such enterprises are ethically justified, that it is not 'cruel' to keep animals under such conditions, chosen essentially for the owners to make money. Evidence is accumulating, however, that frequently animals do suffer under such conditions (e.g. Brambell, 1965; Ekesbo, 1978; Folsch, 1978;). Such arguments are the subject of Chapters 8 and 9 where they are examined.

The argument is also used that intensive husbandry has greatly increased the efficiency of animal-rearing enterprises and therefore provides cheap food for the poor and a 'reasonable income for the producer'. This is a gross mis-statement in terms of food, energy, scarce resource utilisation, economics and social and political systems.

Headlines in the livestock press frequently hail successes in foetal transfer; advertisements for donor or receiver cows are not uncommon. Ecological agriculture does not consider such practices as they are not self-sustaining, use resources and cause unnecessary discomfort and manipulations of animals.

Hormones are not used at all in ecological agriculture, nor any imported food or fodder, so risks of such diseases as BSE are negligible.

Energy

By definition, ecological farming cannot run on energy debts in the same way as modern orthodox farming (e.g. Steinhart and Steinhart 1974; Blaxter, 1976). Fossil fuels are used more sparingly and, in developing countries, often not at all. The chief source of power is provided from the farm, that is, human muscle and animals. This allows for greater employment in the rural sector, and the employee can be providing his own food as well as some income. It would seem a better option than the present choices: starvation, unemployment but with government finance, or employment in the industrial sector with the drift to the cities and the concomitant social, political and environmental problems.

Appropriate technology has begun to develop techniques for using renewable sources of energy such as tidal, wind, solar or water power and methane from manures. The research here has largely been done by private finance. Inevitably the technology lags behind, even though there is one or more of these possible sources of energy available on every farm, depending upon location. The ideas is to develop the extra energy source that is suitable for a particular location. The Chinese farmers, for example, have developed the methane collector to a considerable level of sophistication. Draught animals are also an excellent source of power provided they fit into niches on the farm and produce other products, such as muck, meat, hides, milk products and so on, and in some areas are again beginning to be used. West Africa is just one example. In many parts of Britain, the gusty climate can be harnessed to provide wind power, tidal or water power, and even solar energy can substantially reduce other energy needs.

The application of modern engineering knowledge and skills to develop these other sources of energy to a more efficient level requires investment in research. We have begun experiments (see pp. 155–7).

The Development of Small Industries and/or Crafts to Process the Farm Products

The aim of this approach, as noted in chapter 3, is to process as many products as possible on the farm. Products such as milk should be processed on the farm into cheeses, butter, yogurt and so on. Wheat grown on the farm should be ground and baked and the bread eaten or sold if in excess.

Products such as wool and skins off the farm animals give opportunities to develop crafts such as spinning, weaving, tanning and leather-working. The products can then be used on the farm (for example, harnesses produced from home-grown hides) or sold off for cash income.

Other crafts could develop, depending on the situation. The occurrence of good clay, for example, could give rise to home-produced pottery to serve the farm community as well as for sale. Similarly, basket-making and reed-work, which depend on wetlands for growing the willow and rushes, would develop where appropriate. Other food-stuffs would also be processed on the farm—fruits into preserves, wines and so on. All these products would be primarily for use on the farm, and secondarily for selling to produce cash income to buy objects which are not possible to produce.

Conservation of Wildlife and the Utilisation of Other Farm Resources

The destruction of natural vegetation, hedges and trees and the lack of understanding of humus have resulted in soil erosion in many parts of the world. Less well known is the recent drop in the water table and the growing salinity of this water as a result of the profligate use of water and widespread destruction of trees in, for example, Australia.

The central argument for the conservation of wildlife on the farm is to maintain the ecological balance by increasing species diversity and thereby increasing the stability of the system. It is thus essential to the ecological farming approach. Arguments for conservation of wildlife based on amenity or aesthetic value remain as valid as ever (e.g. Tansley, 1945) but are not likely to have a large-scale effect in poor countries.

On the ecological farm it is not the intention simply to keep rough areas, steep banks or other uncultivable pieces of land for wildlife reserves, laudable though such an approach is. Rather the whole farm, by providing a large number of different vegetation types and niches, is managed to maximise wildlife use and variety in association with introduced animals and plants . . . and humans. As a result of applying no manufactured fertiliser to the grassland, a variety of species will flourish there, including natural nitrogen-fixers such as clover. In addition, an area may be set aside as a woodland, a wet area or moorland, depending on the topography. Hedgerows, shaws, ditches and other such features will also contribute to habitat variety.

In order to achieve this aim without taking land out of agricultural production, various strategies can be adopted. For example, fences can be replaced with hedges by using quick-growing species which rapidly give cover, screening and provide some animal fodder and minerals (Culpepper, 1959; Koepf *et al.*, 1976). These hedges will become more species diverse as they grow older (Pollard *et al.*, 1974). Timber trees can be planted, or left if self-sown when the hedges are trimmed. In this way timber for use on the farm (for burning or construction) will become available. Thus woodland not only provides wildlife habitats, but can also become an important economic asset (see Table 13, p. 119).

In Britain today there are two million acres of neglected coppices (Rackham, 1976). Because such areas are no longer cropped and utilised, the farmer and farm adviser no longer realise their usefulness, nor their monetary potential. There are now beginning to be grants given for small woodlands on farms, but little education on such economically and ecologically sound practices as coppicing. This could well be useful in other areas of the world where there are extensive hardwood forests, such as the eastern United States, Brazil and Australia.

Hedgerows can also be useful sources of fruits and nuts in temperate areas, providing elderberries, blackberries, walnuts, hazelnuts, hips and haws and so on. The same arguments apply in the tropics where food, timber, shelter, fuel and decoration could be supplied from living fences which also act as wildlife reserves. As Schumacher (1974) aptly points out, the industry or harvest does not have to be very large to be judged successful.

The Social and Political Effects of Ecological Agriculture

The adoption of the small unit size and labour-intensive approach of Ecological Agriculture would lead to the employment of more people than in present orthodox high-input agriculture. It is estimated that, using this type of agriculture, approximately two hours per day have to be set aside by an adult to provide food for himself and one child (estimate adapted from modern subsistence farmers and measured off our experimental farms). There are therefore many more hours of the day that can be used for further cultivation of produce for sale, for craft work or leisure pursuits.

This kind of agriculture also has a social effect, and that is an increase in 'job satisfaction'. This results from the variety of work that it is necessary to do to preserve diversity, and also the philosophy of self-reliance. 'Specialists' are not called in for building or plumbing or other jobs. Reduced risks to health and fewer serious accidents are the results of having no dangerous chemicals about, and little if any heavy equipment. Conventional farms today have become dangerous places.

The basis of the farming unit is the family, as in peasant agriculture where it survives today. This may often include the extended family and allows for long-term continuity, particularly important in establishing self-sustaining systems. Necessary cooperation between units at harvest time leads to communal participation, and hence more community feeling. In such a system useful and appropriate work can be found for young and old, male and female, so that a mixed community is maintained instead of the very young or the very old being filtered off into segregated homes.

In much of the developing world such a social system still operates (although rapidly vanishing under pressure), so that in such areas it would involve no large-scale social change. Improvements along the lines suggested in this book could increase yields and provide more varied and nutritious diets than are often grown today in subsistence family communities.

Politically the advantages of such a system are that the majority of farmers can remain independent of government pressure, credit demands and big business, and can thus become a powerful group to defend their own interests. Private ownership is also a great motivator to hard work and to develop that land for which one is responsible, particularly if it can be foreseen that such developments could be continued by, or benefit, one's offspring. Let us not forget that the landscaping and planting of trees in the eighteenth century, which made the British countryside so particularly attractive, was the result of landowners planning for posterity, not for personal profit, although of course this was on a large scale. Experience in Russia, where productivity of privately worked small areas greatly exceeds that from the co-operative, illustrates the importance of the private responsibility for land.

'The proof of the pudding is in the eating', which determines whether or not it is deemed *successful.* The next three chapters describe the degree to which three ecological farms, one in the South-East of England, one in a marginal area, the Hebrides in Scotland, and one in a National Park in Devon, fulfil the criteria of Ecological Agriculture, and tell a little about the adventures we have had and the life we continue to lead.

5 Milton Court: an Ecological Farm in the South-East of England

Producing food is fundamentally a practical business. However well 'qualified' one may be with degrees, jargon and impressive knowledge, all is meaningless until it is done in practice. It became obvious that if I was to question how things were done in agriculture, and my arguments were to carry any weight, then, as a good scientist, I was duty bound to practise an alternative approach. Would Ecological Agriculture really work?

And so, in 1972, we set up Milton Court Ecological Farm, to test the practicality of the system and to develop the ideas. Since the farm was never in receipt of any money for research, one of the tests, and the one that most farmers are primarily interested in, was its economic viability, but it was also important to see to what extent the farm was going to manage to fulfil the other criteria we had set. Were the shortfalls our fault or were they going to be impossible to fulfil in any case? Twenty years on, most people still argue the latter: 'We would love to do it your way, but it's just not possible because of a . . . b . . . c . . .'

How were we going to fulfil the tenets outlined in chapter 3?

1 We must be *self-sustaining* in terms of animal and human food, with no applications of chemicals and minimal inputs from off the farm.
2 We must practise *diversification* of animals and plants, with many species of each. This would result in many different activities happening on the farm all the time.
3 We must aim for *large net yield*, measured perhaps in terms of energy. These energy budgets would then be useful to assess the degree of self-sustainability.
4 We must make use of *existing niches* on the farm and design appropriate ecologically sound and distress-free environments for our animals.
5 We must achieve *economic viability*. Although the agricultural

economic climate was against us at that time, I felt that if there was anything in our approach we must at least cover our costs. Of course, since I had no extra money and very little savings, and no way of persuading the bank manager to lend me money, if the farm failed to cover costs it would just cease to exist . . . natural selection, it's called!

The main argument produced against Ecological Agriculture was that it would not pay. The same argument is still used today, particularly to justify intensive animal husbandry.

6 *Farm size* was one requirement we already met. The farm was substantially smaller than it was considered could be farmed 'economically' in a conventional way without high-input horticulture or intensive animal husbandry. The limit at that time was set at around 30 hectares. Our farm was 13.6 hectares.
7 What were we going to *process* on the farm, and how?
8 How could we integrate *conservation* with utilisation and *aesthetic values* of the English countryside?
9 We must conduct *research*, keep good records and all the figures, so that we would be able to assess the different factors in the future and so have some evidence of our failures and, we hoped, our successes. But could we also discover the areas most in need of research, so that, if there ever were any public money and facilities given to try to answer some of the questions, we would be able to point to the most important questions?

We must remember that the aims of ecological farming are:

a To produce nutritious food in a mixed diet to feed the farm family at no cost other than labour and land.
b If there is land and energy over, to grow cash crops for sale in order to buy those things that the family may require but cannot make, often including, in developing countries, education and some health care.

At the same time, since in industrial nations particularly there is insufficient land and motivation for every family to wish to own and farm a small amount of land, self-sufficiency alone cannot be the aim. The aim of an ecological farm is to *increase the biological efficiency of the system* so that it can produce enough food for many more people than the farm family. We calculated that 0.4 hectares was necessary to produce enough food for a family of four (two adults and two children), and that therefore 16.0 hectares would feed forty similar families. As in our own case, the excess food can be sold to produce a cash income—we used it to pay for labour at conventional agricultural rates.

Unless the developed countries begin to produce their own food from their own resources to a greater extent than at present, rather than relying on the importation of foodstuffs from Third World countries, they may find themselves in a difficult position should the developing countries decide to produce food crops rather than cash crops. The West must put its own house in order before it exports its problems to the developing world. This was why we decided to develop our experimental farm in the Western world rather than elsewhere; the next step will be to develop a demonstration, experimental and educational farm in a developing country with local expertise and labour, to *exchange* techniques and ideas . . . not just to teach! (See chapter 10.)

Milton Court Farm

The 13.6-hectare farm was situated in the Cuckmere valley in Sussex, three miles from the South Coast (see Figure 6). The climate there is relatively mild, with temperatures varying from −6°C to 25°C. The rainfall averages 635mm per year, falling mostly in the winter. This gives rise to flooding in the winter in most years and a dry summer which inhibits grass growth. The Cuckmere valley is formed from run-off from the South Downs (Cretaceous chalk outcrops) which surround the farm. Our soil had a pH of 6.5 to 8 (alkaline) and was deep and alluvial; although it was flooded frequently, it dried out relatively quickly.

Most of the year the river was contained in its levy banks above the level of the surrounding water meadows, but once every two years or so it would overflow and flood right up to the doors of our barns which had been carefully built 120 years previously above the flood tide-mark. The valley was steeped in history, with the remains of an Anglo-Saxon 'motte and bailey' (fortified mound surrounded by defences) ringed by a moat. The moat acted as a drainage channel for that area of the valley. We also hired from a retired literary couple (whom we had persuaded of the importance of our project!) a four-hectare field half a kilometre distant on slightly higher ground, approached by an old, little-used bridleway, lined in spring with crab-apple blossom, and the heady, delicate white beauty of swaying cow parsley.

We kept most of the farm as permanent pasture, but approxi-

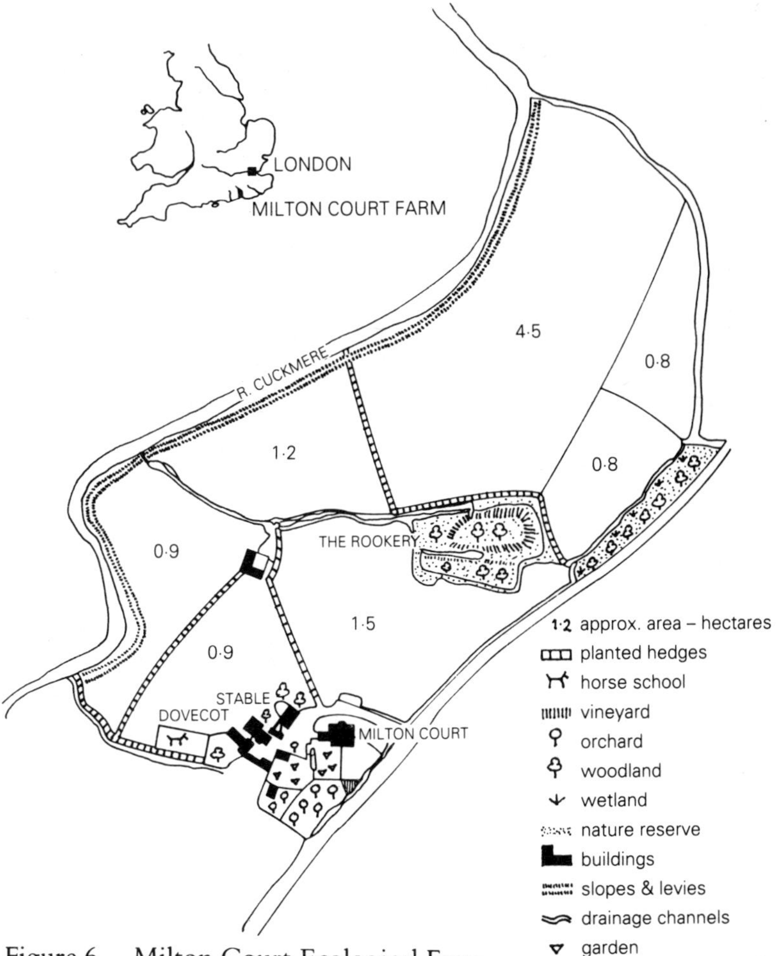

Figure 6. Milton Court Ecological Farm.

mately two hectares were ploughed every year or every other year to grow wheat for our bread, and were then reseeded with mixed grasses. There was one hectare of vegetable and fruit gardens and orchards, many of the trees planted by us in the first winter. The old motte and bailey had been grazed before we arrived, but even so elm trees had managed to establish, and a magnificent bank of small daffodils. A besandalled bearded amateur botanist turned up one day to show us the last remaining wild green hellebore in Sussex nestling by a protective elm trunk. After this, what alternative did we have but to declare the whole area of 2.5 hectares a nature reserve—18 per cent of our farm. The first few years saw the sad death of the elm trees from the spread of Dutch elm disease, despite all our efforts to save them. However, some still remained and now, 15 years later, they are flourishing. As the elm trees went, we planted indigenous hardwood trees that had originally been there in the wild wood days. On such a small acreage, sacrificing an area from grazing like this was hard in the first few years, and it would be many years before wood crops could be taken off it in a sustainable way; however, it did provide wild fruits, nuts and berries with which we made crab-apple and blackberry jams and pickles, hazelnut mushes, and wild garlic salads.

The nature reserve also fed our spirits and delighted our lives by habouring and feeding an ever-accumulating number of wild birds and mammals, and a place for contemplation, where children could camp in the summer, or for ice-skating parties for humans, dogs and ducks in the winter. We kept note of the fauna and flora and how it grew in species over the years (see Table 9). When we left the farm in 1983, we put a covenant on the deeds to ensure that the nature reserve would remain; now, ten years hence, English Heritage wish to declare it a national monument. So fast is history made!

There was a large Victorian house built of flints from the chalk, with walls half-a-metre thick, and a glorious set of nineteenth-century farm buildings, all of which were tumbling down when we took over because the previous farmer had considered them unsuitable for 'modern' farming. They included an old carriage

Table 9. Numbers of species and breeds on the farm in 1982

Fauna:	Wild:	Mammals, 19. Birds, 36. Reptiles and amphibians, 4. Invertebrates, not recorded
	Domestic:	Mammals, 7 (including man); 15 breeds. Birds, 4; 11 breeds. Invertebrates, 1 (bees)
Flora:	Wild:	Angiosperms, 88. Gymnosperms, 8. Non-vascular plants, not recorded
	Domestic:	Vegetables, 44. Fruit, 14. Field crops, 2. Ornamental species, approximately 30. Grasses, 20

Total number of species: 273. Number per hectare: 20.07

house, stables and hay-loft for the working horses, pigsty and run, duck-house, goose-house, granary and cattle yard below, calf-houses, a very ancient and very large dovecot and, remarkably, a chapel, all made of flint with slate roofs. It was a mystery why this working family farm had a chapel; I can't help feeling it might have been for status, since although they were working farmers, the three-storey, 20-room house was clearly divided into the 'family rooms' and the 'servants' quarters'. I have a sneaking suspicion, however, that when none of the local gentry were visiting, they donned their smocks and scampered through the kitchen and out to the farm-yard to milk the cow and feed the horses. We added a dairy and forge built from flints collected off the downs.

The farm was in an area of outstanding natural beauty, heavily protected from any type of development—except for agriculture which is largely exempt from planning controls. But how quickly things change: in the early Seventies we were not allowed to change the granary even to allow accommodation for temporary agricultural labour on the farm. In fact I had to go to appeal to be allowed to have a caravan parked round the back for agricultural labour. Now, twenty years on, each of the buildings has been converted into a chic little cottage with all mod. cons. The yard looks like a suburban garden, or a 'rural museum'.

On my last visit there I could have sworn, as I looked up into the familiar swaying of the two conker trees in the yard, that I saw a crow I knew chuckling; he winked at me and sawed off cackling over the copper beeches.

Even in the early Seventies the population in the area was 100 per square mile, and there was great pressure from weekend visitors from London to allow the land to be used for a wide variety of recreational and educational pursuits. It seemed only sensible and right that we should try to provide for some of their needs and desires on the land that we were lucky enough to have under our stewardship, as well as feed many of them.

Traditionally, for the last few hundred years, the downland around the farm has been farmed as large units. This is not the case in other parts of Sussex where small peasant farms were the order of the day. These large farms are still maintained, the landlords often owning vast tracts of land, sometimes letting them on permanent tenancies as large farms.

THE EXPERIMENT AND ITS RESULTS

So to what degree did we manage to fulfil the aims of the farm? What were the problems and what the strengths of this system?

1 The Self-sustaining System

Initially, we aimed to produce food for people and animals. Within the first three years all the animal fodder was being produced on the farm, with the exception of some hay made from verges and common lands (areas considered too small for other farmers to use). We bought in barley for animal feed as we considered it too risky to grow barley on land which was liable to flood. Since then I have learnt more and realise that we could have grown barley, rye or oats there with little trouble. We also bought in some minerals for the animals in our sixth year, after we had a case of magnesium deficiency in a recently calved cow.

Almost all the food for humans was home-produced, including all the dairy products, and we found out how profitable it was to have a house cow. Ballo the cow saved us £56 a month (at present values over £100), providing a household of between three and ten people with butter, cream, yogurt and hard and soft cheeses—not bad for a little old Guernsey reject from a dairy herd! She was with us until she was 22, never missed a year's calving, taught well over 70 people to milk, and took the micky out of them all.

We had all our own meat, vegetables, fruit, preserves and flour, but we did buy in oil, salt, coffee, tea and sugar. Now and then, when visited by particularly serious purists, we made herbal teas and ground up acorns and dandelion roots for coffee, but the resulting brews were foul, and I decided to suffer their disapproval.

One of the commonest objections to a self-sustaining farm is that the systems fails because one is selling goods off the farm, and therefore will be losing nutrients and trace elements. We were very conscious of this and so took regular soil samples and tests. We expected in particular to find some sign of reduced phosphate (because phosphates are accumulated in animal bones and they are then sold off), but we detected none over this period of time. In fact, as we learnt more about grassland management and farmyard manure and vegetable growing, our yields and carrying capacities off grass continued to grow. I don't think they had reached their zenith when we left. When one thinks about it, however, I suppose the lack of detectable loss over this time is not surprising—after all, 90 per cent of what we were selling was water (animal and plant weight). Nevertheless, it showed how easy it was to retain and even improve the fertility by even rather clumsy management, as ours was in the early days. Nature, it seems, is very happy to co-operate, given half a chance.

The most important method used by ecologists to measure the biological efficiency and production of an area is by looking at the 'energy budgets'—how much energy is used and produced, and how this balances. We decided to try and do this with our agricultural system, and although several others have done this up to a point, we had to make informed guesses here and there. Nevertheless, the energy budgets were much better balanced than for conventional agriculture. We were more energy-sustainable and more energy-efficient than the conventional farms (10:1 and 6:1. See Steinhart and Steinhart, 1974; Blaxter, 1975). We used 2.7 calories for every calorie produced (Table 10), but were still not as good as a natural ecosystem. On later farms we improved (see Table 16, p. 140, and Table 17, p. 158). If we had grown our own barley, this would have made all the difference: 1:2.13 (one used to 2.13 produced).

Livestock and equipment

Initially there was little capital for the purchase of livestock and equipment (Table 11). Gradually over the years the poor-quality stock was replaced by improved genotypes and, with selective breeding, by 1983 we had a herd of South Devon and South Devon cross Guernsey cattle and calves (two each), a flock of 22 Kent and Kent cross ewes, a herd of 12 top-class multi-purpose horses, a flock of 50 assorted poultry, pigs, dogs and so on. The livestock showed a large capital appreciation of 235 per cent which is considerable, even allowing for inflation.

We started with no money so the capital for farm machinery was very small. This allowed only for the purchase of old, second-hand equipment, a small tractor, some hay-making equipment and some horse-drawn equipment badly in need of repair. Additional equipment was subsequently purchased from farm income.

The policy of buying machinery in need of repair, mending it and using it on the farm clearly paid off in terms of its appreciation (66 per cent), again allowing for inflation.

Table 10. Energy produced and consumed on the farm, 1982

a. Animals and people maintained throughout the year

Product	*Individual weights (kg)*	*Individual energy (MJ)**	*Total energy (MJ)*
10 cows	800	9,632	96,320
15 sheep	45	503	7,549
1 sow	120	1,569	1,569
50 hens	2.2	23.9	1,195
5 ducks	3.3	35.9	179
5 geese	4.5	48.8	244
8 horses	800	9,632	77,056
3 dogs	20	216	648
5 people	55	825	4,125
		Total A	188,885 MJ

All weights approximate average: *from Blaxter (1975); some estimates where figures not available.

b. Annual *production* (yield of animals and produce leaving the system in addition to those maintained. Estimated MJ of dead weight)

Product	*Individual weights (kg)*	*Individual energy (MJ)*	*Total energy (MJ)*
15 young beef	400	4,816	72,240
20 lambs (fat)	38	425	8,500
2 pork/baconers	150	1,962	3,924
6,570 eggs	0.056	0.351	2,306
10 hens	2.2	23.9	239
5 ducks	3.3	35.9	180
3 tons vegetables	–	2,700/t	8,100
1 ton wheat	–	17,400/t	17,400
800 gal. milk	–	13.6/gal.	10,880
0.5 ton fruit	–	2,700/t	1,350
1 horse	400	4,816	4,816
		Total B	129,934 MJ

Total produced and maintained (A + B) = 318,819 MJ.

Table 10. (*Contd.*)

c. Annual energy *inputs*

Product	*Quantity*	*Energy/unit (MJ)*★	*Total (MJ)*
Diesel fuel	1,000 l	53.1×10^3/t	42,480
Electricity	4,000 kw	15/kw	60,000
Coal	2 tons	30.5×10^3/t	61,000
Machinery depreciation and repairs	£100	164/£	16,400
Transport	200 ton miles	6/ton mile	1,200
Barley	6 tons	16.7×10^3/t	100,200
Straw	14 tons	16×10^3/t	224,000
		Total C:	505,280 MJ

Total produced (A + B) – Total imported (C) = – 186,461 MJ.
This gives an energy debt of 186,461 MJ, or a ratio of 2.7:1 (which includes all the energy used in the farmhouse). This is disappointing. However, if appropriate land was available (2.08 ha) all the barley and straw could be grown on the farm with little other energy costs.

d. The energy costs would then be:

Diesel fuel	50,000 MJ
Machinery repairs etc.	18,000 MJ
Transport	1,200 MJ
Total:	69,200 MJ

If the energy consumed in the house is omitted (as is normal in such calculations), the operation would then show an energy profit of 249,619 MJ and an energy ratio of 1:3.6 total; or 60,734 MJ *produced* – a ratio of 1:2.13.

e. Other energy produced on the farm per annum

From wood	3 tons at 7×10^3MJ/t	Totalling	21,000 MJ
Solar energy (from solar panels)	900 kwh	Totalling	13,500 MJ
Horse power	5 horse hours/day at 2.5 kwh/day	Totalling	13,500 MJ
Human power	3 people at 1 kwh/day	Totalling	4,500 MJ
		Total:	52,500 MJ

Table 11. Capital investments, 1972–83

Capital investment in livestock (1972 to 1983)

Initial investment (£)		*Investment from farm earnings (£)*	*Present valuation (£)*	*Percentage growth*
Cattle	3,290	500	7,800	237
Horses	1,800	1,500	9,000	500
Sheep	260	60	1,000	384
Poultry	150	75	250	166
Pigs	50	75	100	290
Dogs	50	–	100	200

Capital expenditure on farm machinery

Initial purchases (1972)	*Subsequently bought with farm earnings*		
MF35 tractor	Muck spreader		
Baler	4 wheel trailer		
Tedder (2)	Horse trailer		
Turner	Horse muck cart		
2 wheel trailer	Horse cultivator		
Plough	Phaeton		
Disc harrows	Welder		
Mower	Tools, etc.		
Chain harrows	Acrobat turner		
Spike harrows			
Seed drill			
Tools, etc.			
Total outlay: £500	Total outlay: £1,500	1983 valuation: £3,500	Growth in value: 66%

Capital expenditure on improvements. Including materials but not labour (1972 to 1982 inclusive)

Tree planting	£400	
Hedge planting	£100	
Sheep fencing	£1,500	
Buying coppice and post and rail fencing	£1,500	Expenditure obtained from farm: £4,000

Building improvements and alterations; new roofs, etc.	£2,000	Expenditure obtained from loans: £2,000
New buildings including barn, implement shed, yard, conversions, etc.	£500	
Total:	£6,000	

Building and land improvements
Due to the neglected state of the buildings and land when we took over the farm, many improvements were necessary, including re-roofing additional buildings, fencing, hedge-planting, draining, tree-planting, electricity and water installation, building of solar panels and so on. Home-produced materials and labour were used almost entirely.

2 **Diversity**

Diversity was achieved by having a range of plants and animals produced and sold. The diversity increased over the years, starting with approximately ten species (excluding invertebrates and non-flowering plants) and rising to 273, or 20.07 species per hectare (see Table 9, p. 102). There are no comparable figures for conventional systems, but our figures are likely to be at least double the figures for many of the specialised systems. Gradually the number of domestic plant and animal species and their total numbers grew.

One of the problems of the ecological system is what ratio of land can be used for arable and what for animals if the farm is to remain self-sustaining. I had found (pp. 74–5) that, on European farms run more or less along these lines, around one-third of the area was used for arable at any one time, and two-thirds for animals. Our arable figure was lower: around 15 per cent. The reasons for this were, firstly, that we had flooding land which would be risky for growing many arable crops, except rice—but we did not have the right climate for paddy fields! Another important reason was that we were and are primarily interested in animals, and inevitably how a farm is organised must to a degree take into account the interests and enthusiasms of the farmer . . . if you want her to remain interested and enthusiastic.

The farm produced about 80 tonnes of animal manure annually, which was spread on the grass, but the majority (about 20 tonnes/

hectare/year) was put on the gardens. The human sewage (about 5,000 litres per year) was also pumped out onto the land after collection and decomposition in sewage tanks. The safe recycling of human sewage is something which needs to be much more seriously considered in all countries. Recently, in Western Australia, I was in three separate localities forced off the beaches because of the stink of sewage effluent in an area desperately short of organic waste and water, and where there were no industrial run-offs.

Grassland management and grazing

Although the farm was largely grassland, the grass was treated like a crop and carefully managed. A multi-purpose grazing strategy proved very successful in maintaining and increasing carrying capacity (Kiley-Worthington, 1977).

The central idea was to cater for the selectivity of the different grazing species (cattle, sheep and horses) and to maximise the retention of trace elements and nutrients. The easiest way of doing this, we considered, was by having a multi-species floral layer and grazing it with many grazing species, on the grounds that different species would have different preferences and thus be complementary rather than competitive. A student found that different species and different parts of plants did concentrate different trace elements (Table 7, p. 84). An extension of this work should lead to the ability to calculate the numbers and types of plants required in the grassland to maintain minerals and trace elements.

In order to enhance growth it is important to retain high levels of nitrates in the soil. These salts are the building blocks of cells. Clovers and other members of the pea family fix nitrogen from the air with the help of special root bacteria. The addition of farmyard manure increases the nitrates, and also ensures high levels of potassium, another important nutrient that is usually applied to conventional farms by manufactured fertiliser.

One very important factor in the management of grazing animals is how to control parasites picked up from the grass. By grazing the one-hectare fields in rotation and cutting each field once a year for hay, by harrowing the mucks and rolling the fields, we managed to cut down risk of parasitic re-infection.

In this way we developed and improved the management of our grassland so that it had a very high carrying capacity (3.6 bovine units per hectare, which included winter feed) . . . higher than some of our neighbours who used up to eight hundredweight of nutrients (nitrogen, phosphate and potassium) per acre per year.

3 Net Yields

We have already mentioned that what we had to do was assess the net yield rather than the gross yield. Table 12 shows what these were, and 1,626 kg per hectare is not a negligible figure in this part of the world. The most interesting point was that, because the inputs were minimal, the yields were relatively high although not comparable with conventional farms' gross yields. It would be interesting to compare *net* yields from a conventional farm but this has yet to be done.

The carrying capacity of both the grassland and the garden was high. The garden also produced a large net yield, and it was interesting how little land was needed to support one family in the

Table 12. Net yields from Milton Court, 1982

Grassland carrying capacity	3.6 bovine units/ha including conservation	
Wheat:	2.4 t/ha	
Orchards:	3 t/ha	
Edible/saleable: biomass from garden:	8 t/ha	
Milk:	33,341 litres per annum	
Beef:	15 stores at 400 kg	Total: 6,000 kg
Lamb:	20 lambs at 40 kg	Total: 800 kg
Pork:	2 baconers at 150 kg	Total: 300 kg
Eggs and poultry:	15 kg poultry meat per annum	and 4,800 eggs
Human biomass supported:	6 at 60 kg; 1 at 12 kg	Total: 372 kg
Dogs (maintained and produced):	4 at 10 kg	Total: 40 kg
Total biomass of animals supported, including humans:		20,490 kg
Total biomass of animals, including humans, supported and produced per hectare (excluding wild animals and the nature reserve):		1,626.1 kg

All the yields are net yields. Comparative figures for net yields from conventional farms in the neighbourhood are not yet available. However, the carrying capacity of grass averages approximately 2.2 bovine units per hectare when applying 240 kg of NPK fertilizer per hectare per annum.

climatic and environmental conditions associated with the farm: 0.62 hectares (1.5 acres) were sufficient for two adults and two children when meat and dairy products were included as part of the diet.

4 The Concept of Niche and Ethologically Sound Environments for Animals

The animals grazed were cattle, sheep and horses. These species were run with males throughout the greater part of the year and were maintained in the kind of group structures in which they have evolved. The young were raised by the mothers and there was no weaning except when the animals were sold off (for example, for fat lamb). The cattle and horses were kept indoors in groups for about four months in winter to prevent 'poaching' (spoiling with their feet) of the wet pastures and subsequent retardation of grass growth in the spring. The sheep were out at grass all the year and lambed outside with a minimum of management or interference. The ram was run with the ewes throughout the year in most years, hence lambing was relatively early (January to February); normally the rams are only allowed to run with the ewes for two to three months per year in order to try and synchronise lambing, and the modern trend is to lamb indoors. This often results in a great deal of work and expense for the farmer and many behavioural and often physiological problems for the ewes and lambs.

All grazing animals can convert cellulose (from plant material) to animal protein and thus do not need to be fed food which is high in protein, although this has become current recommended practice on farms. If they are fed high-protein foods these are usually bought in by the farmer and may include fish meal, soya bean meal, dried milk, and ground-up waste meat and faeces of animal protein. This can cause a whole galaxy of environmental economic and ethical problems in other parts of the world, as well as spreading disease such as BSE. One can't help thinking that the establishment's advice and research efforts to try in this way to grow animals faster and bigger and to make more money is particularly silly! Our animals were fed only home-produced hay, straw and small amounts of barley at crucial times (for example, around parturition for two to three weeks). No hormones of any type were given. These are often used to synchronise oestrus in ewes so that lambing comes all at once. Nor were any drugs used to begin labour at convenient times (they are quite often used on humans to ensure that birthing will take place at 'social' hours).

Beef production

During this period we conducted research into how and when a cow recognised her calf, and developed a method of adopting second calves onto newly calved cows so that they would think they had twins (Kiley-Worthington and de la Plain, 1983). The idea behind this was to ensure that more calves were mother-raised outside rather than in crates on buckets and to increase the financial respectability of the suckler cattle. Eventually we had a double-suckling herd of South Devon cattle which still continues to grow in quality if not in size.

Dairying

Like the other enterprises on the farm, the dairying was small but profitable. One or two cows were hand-milked. The milk was processed into cream, butter, soft cheese, a 'Port Salut' cheese, Gouda and Cheddar-type hard cheeses, and yogurt, and sold retail through the farm shop. In addition, the cows provided all the dairy products required for the house.

I did not know about cheese production to start with and so I sent a young woman student on the farm on a weekend course to learn all about small-scale cheese production. She came back with reams of typed foolscap pages telling her how to make a cheddar. She then spent one whole day making a 3lb cheese which, when we ate it, was perfectly horrible, not because she was incompetent but because the training was. The mumbo-jumbo surrounding even something as simple as home cheese-making is remarkable. It was enough to put anyone off trying, and I guess it did. After she left I gradually cut out almost every step until now we have several very successful recipes which take all of ten minutes with the milk straight from the cow . . . and almost no equipment.

We had more trouble: qualifying for a licence to sell dairy products and raw milk, and it looked initially as if it would really not be possible without spending much more money than we would ever be likely to recover. However, cussedness or perseverance eventually reaped rewards: the dairy inspectors changed their attitude from complete dismissal to helpful co-operation and the great day arrived when we obtained our full dairy and raw milk licence.

Sheep

Our first sheep were a few reject ewes bought for £1 each at the local market. Surprisingly they produced lambs, and gradually we acquired our herd of Kents, a big lowland sheep. We crossed them with Hampshire Down or Suffolk ram and had a good lambing percentage, around 150 per cent. We sold some of the fleeces to local spinners, but most of the wool went to the monopoly wool marketing board.

Horses

The horses were (and still are) our working partners, friends, research subjects (their behaviour is studied) and money earners. They did appropriate jobs on the farm (harrowing, bringing in the hay while the tractor baled, and so on). They provided us with fun and recreation when we did competitions, holidays and treks, they taught people, and they bred. We had visiting mares to our Anglo-Arab stallion and we sold our own youngsters. They seldom lived in single stables, we did not wean the youngsters and the stallion ran with the mares so that he could court and they could mate as they wished. This has become very unusual practice in horse breeding (see Kiley-Worthington, 1987 for further discussion).

Pigs and poultry

The pigs were used as scavenging omnivores, and they were fed barley, brewers' grains, whey from cheese-making and scraps from the house and garden. They dug the garden by rooting during the winter, saving us that chore, and they bred. We sold the meat retail, and made hams and sausages.

The poultry were also treated as scavengers and were fed the same diet as the pigs. They were run free in the orchard and, with the pigs, were used to go over the garden to remove pests and to scratch and dig during the winter months. Egg production was lower than in the conventional system on this diet but the net cost of the eggs was much less. Cocks always lived with the flock and all the poultry were encouraged to lay and sit on their own eggs, and to look after their own offspring. Modern hens have been selected not to go broody and are extremely bad at sitting on their own eggs and raising chicks. We often had to borrow or buy bantam hens to raise clutches, but over the years we selected for hens who would do this well, and from frustrating beginnings (a hen would suddenly stop sitting when only a week away from hatching, and one would find all the eggs cold when one went to feed them in the morning) we ended up with good layers and good mothers.

After one Christmas spent plucking, breathing, cursing and spitting turkey feathers for a whole week, I vowed never to raise free range turkeys again . . . unless one could get the consumer to do her own plucking (which one never can!). Geese are grazers and could have been fitted into our grazing system in large numbers, but the same problem did not fill me with enthusiasm, even if it would have made lots of money. We had a few geese, though, and they were excellent watchdogs, controllers of infant trespass, and dog discipliners, as well as grazing otherwise little-used areas such as the lawns and the orchard, if they wished . . . one has to be polite to geese.

Animal and plant diseases

It has often been suggested that balanced ecosystems are more stable and should suffer less from epidemic diseases (e.g. Odum, 1971). Thus fewer prophylactic drugs or treatments should need to be used in a balanced farming ecosystem, and at the same time there should be an increased longevity and net production of the stock. Although there were various pests and diseases in the market garden, these never reached epidemic proportions or caused economic loss of any importance. No organic potions or inorganic remedies were used against pests or diseases. Techniques such as companion planting, rotations and the use of aromatic plants were, however, used in an attempt to reduce infective and destructive organisms.

We encountered no serious disease problems in any of the livestock, although there were the usual accidents. For the first six years prophylactic worm treatments were used. However, after a case of fluke in a ewe and a worm-induced peritonitis in a foal, regular prophylactic worm treatment was given to the sheep and to the young cattle and horses. No doubt there was an inevitable increase in parasite numbers as a result of intensive stocking over the years. Some preventative medicine was used and, where necessary to relieve suffering, antibiotics and surgery, but these were kept to an absolute minimum.

Longevity
One of the advantages of a low-input system, where the livestock are not being 'forced' into high production levels by high feeding rates, early maturity, high production and so on, is that the longevity appears to increase. The farm would really have needed to continue for another 30 years to accumulate enough figures on this, but we still have many of the same animals ten years and two farms on. One mare was bred until she was 25 years old and two cows were still in the herd at their 14th lactation, never having missed a year of breeding. Thus disease problems were relatively few and there is a possibility that longevity was increased under this system.

5 Economic Returns and Viability

All aspects of the farm showed a rise in profit over the ten-year period, with the exception of poultry which proved to be one of the more difficult industries to keep economically viable, despite the very ready market for free range eggs (see Figure 7).

The main point of interest is that the costs of each enterprise were very low, as a result of feeding home-produced food, housing in existing buildings and low veterinary bills. However, the overall profits were not very high (Figure 8), although they were very much better than would be expected with this size of unit in this part of Britain. The labour charges were relatively high because of the number of people employed per hectare (one to 7.7 hectares). The system was labour-intensive and could support one-and-a-half professional workers, although it was considered impossible locally to support even one worker off an area of 41 hectares (three times the area of our farm) unless intensive, high-input horticulture or animal husbandry was practised (personal communication from the Ministry of Agriculture, Agricultural Advisory Service representative for East Sussex; University of Reading Farm Business Data, 1982). So although the returns per hectare were not very high, when combined with all the other attributes of the system and the high employment, it did show some promise as a viable economic alternative (see Table 13).

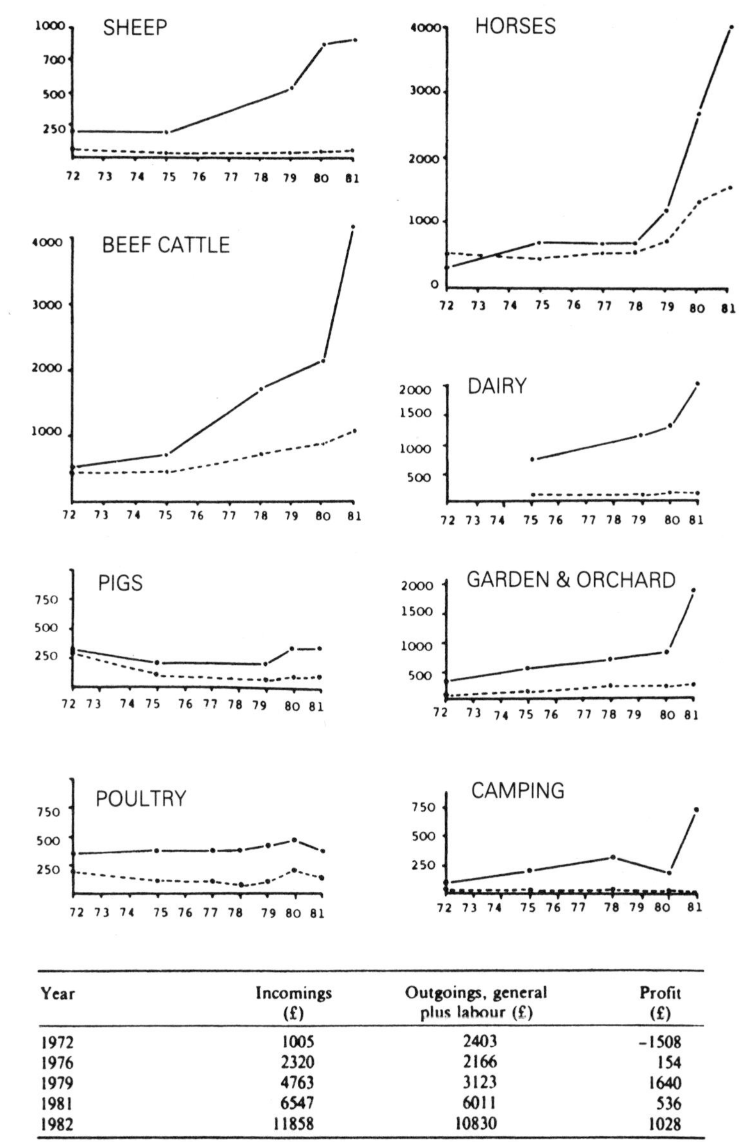

Year	Incomings (£)	Outgoings, general plus labour (£)	Profit (£)
1972	1005	2403	-1508
1976	2320	2166	154
1979	4763	3123	1640
1981	6547	6011	536
1982	11858	10830	1028

Profits per hectare: £64.5 (excluding charge on land capital).
Area for each worker: 7.7 hectares.

All figures in £ sterling. (———— income; ---- expenditure).

Figure 7. Income and expenditure at Milton Court, 1972–81.

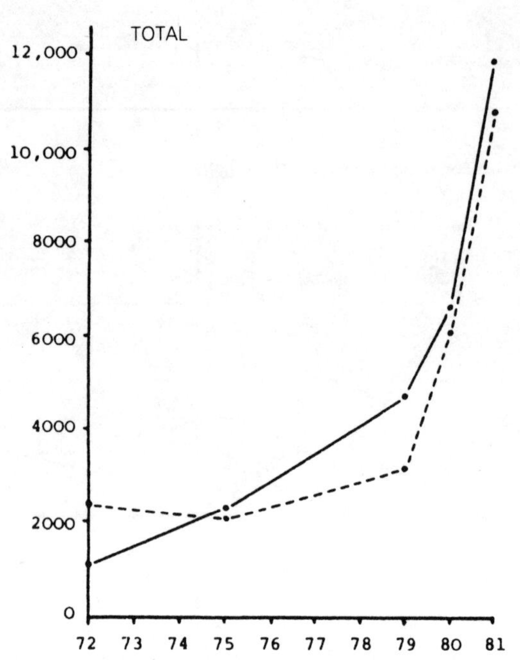

Figure 8. Total economic returns for the farm between 1972 and 1981. All figures in £ sterling. (——— income; --- expenditure).

We did not take subsidies for improvements, fencing, draining, building, and so on. The only government subsidies that were accepted were a beef cow subsidy in 1981–2, EEC intervention payments for fat lamb in 1981–2 and a beef calf subsidy in 1974–8. A small tree-planting grant was also accepted in 1975. The reason for not applying for state grants for improvements was to test if such a system could be independently economically viable, rather than becoming a subsidised service like conventional agriculture.

Another point of interest was the large increase in the capital value of the livestock and even of the equipment. This was due to the policy of low capitalisation and then breeding superior livestock and repairing and restoring old machinery, which is often more appropriate for this system. Thus the initial capital investment in running stock for the farm was relatively low (see Table 11, p. 108), but capital growth was high: 66 per cent!

Table 13. The economics of Milton Court for 1979–80

	Total incomings:	*Total outgoings:*
Beef cattle	£1,449.38	£265.00
dairy	£222.93	£5.10
Sheep, lambs and wool	£584.47	£7.70
Horse and riding	£1,147.00	£88.20
Pigs	£51.37	£75.00 (bought)
Poultry	£147.34	£64.48
Vegetables and fruit	£54.15	£36.81
Camping	£275.00	
Farm general	£584.47	£1,699.25
Provision of food/accommodation, etc. for labour	£3,800.00	£3,800.00
Cash payments		£1,140.00
Total	£8,316.03	£7,181.54
Profit	£1,134.49	
Gross incomings/ha:	£611.47	Number hectares/ man: 10
Profit per hectare:	£83.41	10

Capital growth in stock from 1972 to 1980: 200.4% = 25.05% per/annum

The capital cost of buying the land was high, and the rents of the land have not been calculated into the equation since land in this area of Britain, within commuting distance of London and three miles from the south coast, was so costly that it was not expected to yield rent from any agricultural enterprise; its value for amenity, beauty and residential purposes quite outstripped its agricultural value. In November 1983 the value of the farm was approximately £15,360 per hectare.

Labour

An agricultural system such as ours was by definition more labour-demanding than the highly capitalised agricultural strategy current in Britain, while the total capital investment in machinery was relatively small—£2,000. Because of the demand from young people to learn more about this type of farming, the farm also had to take on an educational role. Students came to learn and repay in part their tuition and keep by working, although the cost of their maintenance and education was often greater than the work they produced. Still, they also gave us a lot of social enjoyment and fun. From time to time, however, the farm was run efficiently and effectively, solely by one-and-a-half professional workers. It also acted as a host farm for various volunteer groups.

Self-reliance and use of materials taken from the farm for construction and repair were central to our aims, so workers had a very wide range of experience and no job monotony. Judging by the number of volunteers who made themselves available, this was a particular attraction of the system. In the course of one year we had six students working for more than one month, approximately 25 on visits of less than one month and about 600 day visitors. Running an 'open house' farm of this type was an educational experience for us, too. I was never really interested in the human social aspect of the farm, or in trying to establish communities, which many idealistic people were attempting at the time. I never did want to live with many people from many different backgrounds and cultures, some with post-doctoral education, some having left school hardly able to read and write—but that is what I ended up doing. I guess that I learnt and continue to learn as much from all these people on how to live and work together as I teach them . . . but nevertheless, I still learn more from the animals I live with! In retrospect we had many hilarious and outrageous times . . . an ever-accumulating fund of stories to amuse, amaze, shock and delight.

Management and appropriate technology

The disadvantage of the system mainly related to management of the various enterprises. Such a diverse operation, in which many small schemes were running concurrently, was managerially very complex, and indeed not everyone would wish to farm this way. However, such a system always remained challenging and interesting, so that the motivation and energy to keep going was always there, even in the roughest weather with food at its scarcest and broken-down machines. It helped to have others there, to have to encourage and motivate, to think always positively and above all laugh a lot.

It is often said that 'organic' farms succeed only because they are run by well-motivated farmers. This may well be true, but the same applies, surely, to any agricultural system, or to 'work' in general.

The system was designed to be more labour-demanding. In the developed world it is often easier and cheaper to buy another piece of equipment than to employ people who have ever-increasing expectations, in terms of both money and job satisfaction, and this is a temptation that management has to face from time to time. There is, however, no shortage of students from all over the world, who are interested in learning about this type of agriculture and in working on a farm where the approach is more holistic and where types of job are very varied and skilled. Only one in four of the applicants could be catered for. There is also a rising number of unemployed in the developed countries at present, and thus it would seem to be appropriate to develop a higher labour-demanding agriculture. However, as Blaxter (1975) noted, there is seldom any real decline in the number of people required to work in agriculture in a particular area, since if they are not working directly on the land they are working in factories producing machines and materials to use on the land. To date, resource scarcities suggest that in the long term it may be necessary for people to be employed in direct food production, thus using fewer resources, rather than indirectly by manufacturing machines for use on the land. Another aspect that has to be considered, particularly in relation to this type of agriculture, is the use of appropriate machinery. Given that capitalisation must be low, is it appropriate to use any machine, or fossil fuel, and if so, which? These questions can only be answered on individual farms, depending upon their facilities. On our experimental farm we always had to keep in mind that labour costs were high, and that if a job was to be done by hand, or with small tools, the cost of doing it might be greater than the value of the result. Thus it was necessary to decide which energy sources and machines were appropriate for which jobs.

To take an example, it would usually be inappropriate to use the horses for ploughing since they were about eight times slower than the tractor, except at certain times of the year when, because of the wet, only horses could plough . . . The tractor got stuck, and pulling it out with the horses was a long and frustrating job, to be avoided if possible. Of course, the tractor was invaluable for doing things that only it could do, such as using the hydraulic lift or baling. On the other hand, the horses could often be an extra source of power, such as during hay-making, bringing in the hay while the tractor baled it.

Organically grown products

There was then, as now, a growing market for fresh, organically grown products which had not been treated with fertilisers and pesticides, or raised in artificial conditions and fed prophylactic chemicals and growth promoters. The techniques of organic cultivation of vegetables and fruit are relatively well understood. The aim of the garden was to obtain the optimum sustainable yield and to see how it compared to conventional techniques. Catch cropping, close planting, interplanting and companion planting were used to achieve this aim. In this way 0.2 hectares yielded fruit, vegetables and potatoes for six people and for sale. The large and varied diet produced in this way for the workers on the farm meant that, over the last six years, the only items bought in, as I have already mentioned, were coffee, tea, oil, sugar and a few luxuries. We need not, of course, have bought in these things, but we had decided that there were some luxuries we would allow ourselves, so long as they came from the right places. At least we knew we were making compromises!

The garden needed to produce a variety of crops to increase diversity and so guard against specific crop failure. For the last four years 12 crop varieties were available at Christmas and up to twenty in the summer. Pest and disease attacks were minimal and no preparations were used against them. The approach was to develop a balanced ecosystem rather than rely on curatives and preventative pest and disease control.

Marketing

The majority of the produce was sold through the farm shop which developed a well-established trade. The farm was inspected and approved to use the Organic Produce Symbol (by the Soil Association, UK) to sell dairy products, meat, fruit, vegetables and grain. No premium was charged for these sought-after products when compared with what was sold in local shops. It was the policy not to extract premiums for organically grown products although the market could have easily supported this.

6 **Processing of Farm Products**

We built our small shop in our fifth year and, apart from fruit and vegetables, sold meat and meat products such as beef, lamb, pork, bacon, ham and sausages, poultry, eggs and a variety of dairy products including cheeses, cream and yogurt, all made and produced on the farm. We also had some locally made crafts such as sachets of herbs, our own pottery, wrought ironwork and leather goods, made by our own blacksmith and leather worker, macramé made from baler twine, and other home-made items.

The development of the farm craft centre was erratic, but during our time there we provided studio space for the leather worker (working our own hides), for spinners and weavers, a potter (clay from the river) and facilities for wood-work from home-produced woods, basket-making from home-produced willows, harness and rope-making. We built a forge and did all the welding and blacksmithing, including shoeing, but we did buy in the iron.

7 Integration of Conservation and Aesthetic Considerations with Utilisation and Production

Since no pesticides, herbicides, fertilisers and so on were used on the farm, and since it was managed in small paddocks where, in general, permanent pastures were maintained, the whole farm was considered a conservation area. We bought some chestnut coppice that had been cut ten years previously and was ready to cut again, and worked it to fence the farm boundary and all the paddocks with the traditional split chestnut post and rail, tongue and groove fencing (see Table 14). We planted hedges behind the fences everywhere where we could ensure both sides were protected. We did not dig out the waterways to improve drainage as this would have resulted in the loss of some natural plants that we were encouraging, for example the kingcup (*Caltha palustris*).

One agricultural adviser was asked, as a test, how we could maintain the drainage without the prevalent scorched waterway policy—much to his consternation. He returned, however, with interesting suggestions, such as digging a ditch to the side of the main drain to prevent deterioration in the drainage of the grassland. This we did and it worked well.

The nature reserve area was planted with approximately 400 trees, including a mixture indigenous to chalkland and some commercial softwood trees planted to act as a nurse crop and to allow for some early cropping. About 0.1 hectare was planted with sweet chestnut (*Castanea sativa*) for coppicing later. This area we calculated would be sufficient to keep us in fencing replacement, and other building and rough furniture-making needs. Cricket bat wil-

Table 14. Profit from buying and working a chestnut coppice near Milton Court

Cost		*Production*	
Labour: at £10/man/day = £50 × 5 =	£250	Produced: 3 benches at £25 =	£75
Transport =	£50	Fencing 500 m at £3/m =	£1,500
Cost of coppice =	£30	Gates/hurdles at £25 =	£50
Labour for fence etc. =	£300	Building materials for reconstruction =	£30
Total	£630		£1,655
			£630
			£1,025

The net gain was £1,025. If more farmers realised the value of their timber to themselves for farm work, perhaps less woodland would be grubbed up or neglected.

lows and other willows for basketry were planted in appropriate locations. For the rest, the natural vegetation was allowed to regenerate and we gathered fruits and berries from it, as well as other plants according to the season (blackberries, hips, haws, elderberries, cow parsley, young nettles, walnuts, chestnuts and so on).

8 **Research**

There are plenty of private foundations and individuals who are doing research. In the last year I have had five questionnaires to fill in for 'researchers' with government money, 'surveying' organic farms. The one thing organic farms *don't* need is more surveys! They need answers to the questions with which they have bombarded the establishment but which have been ignored. There is little likelihood that the situation will change until most of the current research personnel have retired, since their training has not fitted them to ask the appropriate questions, or to use the necessary multi-disciplinary approach in trying to answer them. For what it is worth, I have listed in an appendix the most important research questions that have arisen from our experimental work.

* * *

Having retained the traditional buildings of the region, flint with timber frames and slate or tiled roofs, the resulting farm-yard, planted with trees and having some animals free range in the yard, others mother-raised in the fields, contrasted strongly with neighbouring 'agribusiness' holdings.

It was a very pleasant place and on our annual open day in August (which started as one of the first fêtes ever to raise money for the 'Ecology' party as it was called), people came in their hundreds to visit, be entertained and learn. One year we had bought some standing hay on the top of the downs with a magnificent view to local roads, only to find it was full of ragwort (*Senecio jacobaea*), a poisonous herb in hay. Instead of buying the grass we should in fact have been paid to cut it! We spent three or four hot days pulling ragwort on the top until time ran out and we decided to abandon our money and the grass. Before doing so Chris, my partner, decided to cut a swath in the hillside advertising our open day. It came out beautifully. It continues to make me smile when I think of the tenant of the land (who had refused to repay our money), his teeth grinding with rage.

We had a great attendance at our open day and I think everyone enjoyed the farm walk, trying to milk the cow, watching the horses dance, playing 'place the penny' with the dogs and drinking tea and eating scones in the garden.

Why did we leave it all? Adventure and research called. What is life, after all, but an obstacle race in which one seems to graduate to the harder, more difficult challenges as time goes on . . . Is this what is called accruing knowledge, or merely stupidity? We did not know or care.

6 The Druimghigha Project on a Scottish Island

In 1983 we had completed ten years on our test study farm at Milton Court. In this part of Britain the winters are relatively short, the sun shines much of the time, the rain falls when it is needed, the soils are deep and fertile, and the relaxing, lush atmosphere of the English countryside prevails. Although the commuters moan about the railways and the roads, by international standards they are superb, and distances short. If you have to drive more than five miles to the station, that is an expedition. Markets are on the doorstep, equipment and spares of every type available within a few miles, auctions every day of the week within a radius of 50 miles. Compared with much of the world, the prevailing standard of living is phenomenally high.

Biologically speaking, it is a rich and productive area. As one wanders around the meadows on a summer's day, the benign smile of the healthy, well-fed animals and plants meets one at every glance. Producing enough in the summer to feed animals and humans through the winter is no problem; mounds of apples and buluses (wild hedgerow plums) rot on the ground unharvested, barns creak with the stacks of hay. Straw is so plentiful that it becomes an embarrassment to many farmers, who burn it. From the end of July the combines grind around the fields, pouring grain by the ton into over-full grain silos. In such an environment in Africa, everyone would relax, lie around, play and sleep . . .

The criticism of the results of Milton Court Farm revolved mainly around this. The environment was too easy: the climate too benign, the soils too fertile, the people too rich, the infrastructure too well organised for the results to be of any interest to the majority of people who starve and live in areas often called 'marginal'. The meaning of this word is obscure, but it is usually taken to mean that the growing of food in such areas is difficult and may not always work out; less 'biologically productive' might be a better term.

It seemed to us a valid criticism, and we determined to see if this were the case. The next six years were to be devoted to starting again and trying out Ecological Agriculture, as we had defined it, in a marginal area. The problem was, where? Africa was tempting, but the veterinary restrictions on the movement of our animals (and we were not going to leave our animals behind) were daunting. The political insecurity of many countries also had to be borne in mind; even if we did not worry, the bank manager (from whom we now borrowed much money) certainly would.

There is plenty of marginal land in the United States and Canada, so I arranged a lecture tour around those countries to have a look. Australia was a further possibility, so we went for another lecture tour, and managed to have a detailed look at several areas. Again the veterinary restrictions and expense were prohibitive. It looked as if it had to be in Europe. France, even parts of Germany, were considered, but finally the most sensible place seemed to be somewhere in Britain, perhaps in the marginal hill land where there had been agricultural and social problems for a long time.

After endless trips on the overnight bus to Glasgow and Edinburgh, looking at many farms, wrestling with the obscure Scottish legal system for buying property, we managed to find an Englishman who would discuss price, and bought a 400-acre hill farm. It had no land that had been used for arable or even any improved grassland. The potential arable land was marked by topography and the presence of bracken (*Pteridium aquilinum*) which denotes a relatively deep and reasonably drained if acid soil. The farm ran from sea level, bordering on a sea loch, to 600-plus feet above (200 metres), and the majority of it was bog (Figure 9). It had a ring fence and a comfortable house, although the farm buildings were minimal, and their further construction later proved to be one of the many major obstacles that had to be overcome. One of the very attractive aspects of the farm was the large amount of woodland—about 100 acres of birch and hazel, and even the odd ash tree still left. The reason for this, in an area where trees are now almost totally absent, was that the previous-but-one owner had been a flower gardener, and had had practically no sheep. As a result, the trees had managed to survive and some woodland had even rejuvenated.

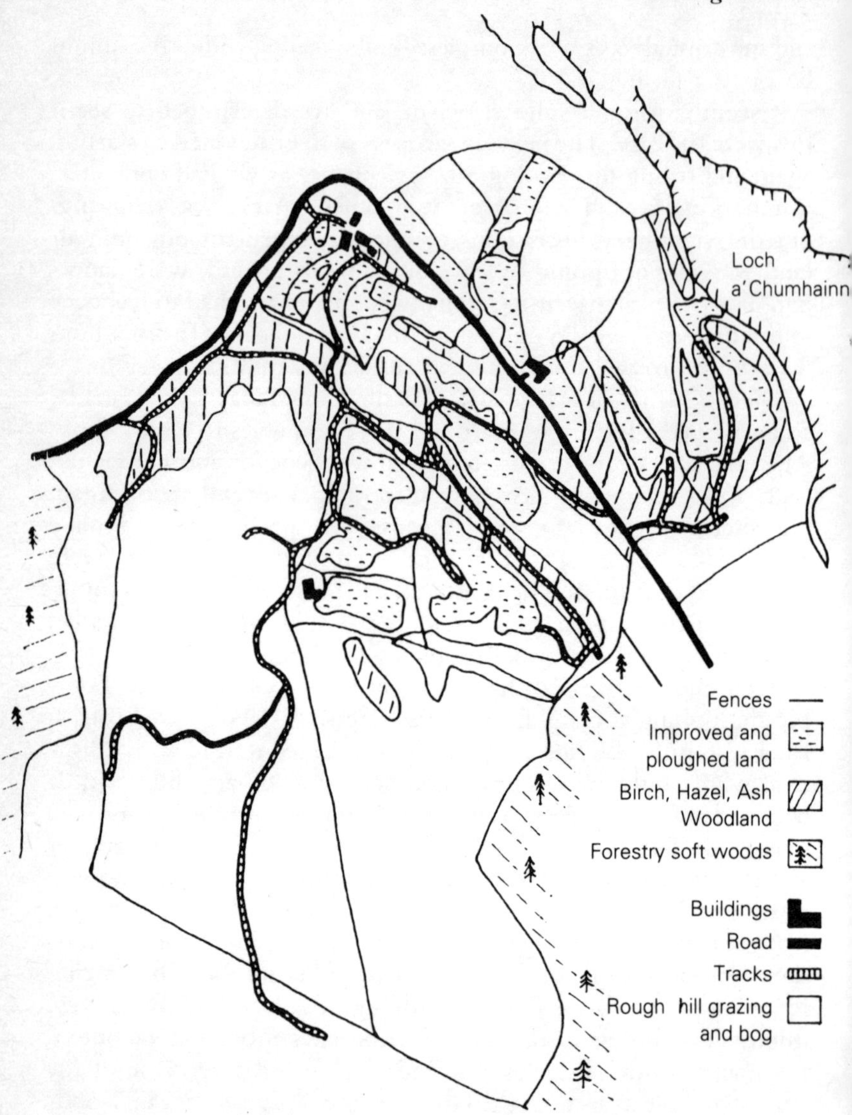

Figure 9. Druimghigha Farm, 1988. Scale 1:10,000.

In addition, the farm was on an island, the island of Mull in the Hebrides, although only a mile (two kilometres) from the mainland and in an area where populations were low and depressed. Because they live in a welfare state they are not starving but almost all are supported by government grants of one sort or another. There was no doubt about it, the area was 'marginal', and had the added difficulty of being an island.

It was not a dry desert, but it certainly was a wet desert, and as man-made as much of the desert lands in developing countries. The difference here, however, was that the desert had been made by the greed of individuals. The land tenure is still intensely feudal, with a few owning enormous acreages, often as investments. As a result, small acreages on which to try and grow anything but sheep are difficult to come by. In marginal areas of the developing world, though, the land is often communally owned, and the pressures have been for mere survival, not for riches.

In October 1983 we moved stock and barrel (no locks!)—furniture, tools, horse-drawn equipment, students and all—the 800 miles (1,000 kilometres) from Sussex to Mull (Figure 10). We went from late summer in the south to what seemed like mid-winter, and it rained continuously for six weeks after we arrived! The

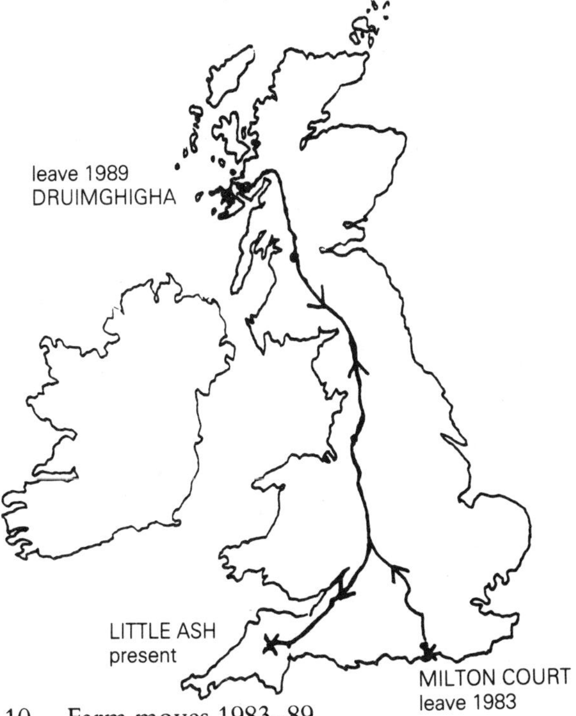

Figure 10. Farm moves 1983–89.

animals still had their summer coats, the grass had all been eaten off by the previous occupier's stock, there was no shelter and no extra food in store. We had no tractor, and not much other equipment. The saga of the move is another story. We crashed a lorry loaded with peacocks, horses, dogs and students, and a full deep freeze with all our winter's food which spilt and melted in the rain over a highland road.

We had to some extent bargained for extreme financial worries (we had a £50,000 overdraft), the weather and other physical difficulties; but we had *not* bargained for the Calvinistic culture. Enjoyment of the present life is considered a sin, as it is meant to be a trial to prepare one for the Kingdom of God. As a result, positive thinking is not very common. This was not a useful approach, particularly when one was taking on an ambitious project fraught with difficulties.

After a year or two I realised what an enormous influence my African background made, and how extremely important early cultural experiences are to general attitudes! The African cultural attitude, unsullied by Western belief systems (as I interpret it, anyway), is that you may be cold, hungry, poor, generally with troubles, but you are alive, and life is to be enjoyed. I well remember one cold morning in the Kenya highlands driving in the freezing dawn mist towards Nairobi, and seeing a young man with hardly a stitch on, a stick in his hand and a few scraggy sheep grazing on the non-grass at the side of the road. He was dancing, and as I passed in my warm motorcar he smiled and waved. He had nothing, he was cold, he had probably had nothing but a small share of *posho* (maize meal) to eat in the last 24 hours: what had he to be so happy about . . . ? He was alive, the air was free and the sun would come up . . . His sheep danced here and there too, and they seemed to have even less!

The cultural influence may be the crucial factor controlling agricultural development in Scotland. It is in such environments that total physical and mental self-reliance must become a full-time reality if one is going to survive and remain relatively sane.

The first winter was a time when all of us, human and animal alike, were struggling physically and mentally; we were walking constantly along the edge of a precipice between survival and extinction, both mental and physical. It was one of those times that one looks back on and perhaps sees it as a necessary part of the experience of life, but every morning I still wake up and thank the world that it is not that time again!

Our standard of living dropped, and so did the animals', through the winter, and I think we are entitled to regard many of the problems we encountered as similar to those of the poor and exhausted, ill-fed and desperate of the developing world. I think we do now have more real understanding and personal practical experience of how to survive—and how one might not. We also have a great knowledge of how little there is to eat in the hedgerows in the winter in the Hebrides! However, we all managed to pull through. That was ten years ago now, and the experiment is completed. We have found out the strengths and weaknesses of the approach in a marginal area, and we have learnt much about techniques and, perhaps most important of all, mental attitudes.

Over the last 150 years the Highlands and Islands of Scotland have become biologically down-graded and less productive, largely due to human mismanagement. In this the area has many parallels with the Sahara's southern border. The sheep have done to the Highlands and Islands what the goat has done for the southern extension of the Sahara. Our aim was to find out if it was possible to reverse this biological downward trend and set the farm on an ecological recovery curve.

History of the Area

The north end of the Isle of Mull, where Druimghigha is situated, is on predominantly basaltic tertiary lava which results in the terraced appearance of much of the landscape. Because of the North Atlantic Drift the climate is considered milder than other areas at the same latitude, with temperatures ranging from −15°C in winter to 22°C occasionally in July and August. The winds and chill factor offer a challenge for the grower in creating and maintaining micro-climates for his plants and animals. The area receives around 1,500mm of rain annually.

The indigenous vegetation types of this area include open park-like forests with a rich ground flora, supporting a great variety of species depending primarily on the drainage and depth of soil. However, little of it remains anywhere in Scotland.

We found deciduous woodland in the valleys and more protected areas, including some 100 acres at Druimghigha. Blanket bogs are found on the impermeable flat tops of the lava flows. These areas are particularly interesting to ecologists, with their complex ecology in a very acid environment. The heather-dominated moorlands originally occurred above the tree line or in exposed places, but now, with the destruction of the trees, they have spread more widely (Pearsall, 1965). This, coupled with a burning policy, has resulted in large areas becoming dominated by bracken, a sign of improvable ground if nothing else; it might be possible to bring such land into arable production.

The island of Mull has been inhabited for a very long time. By the nineteenth century it was supporting a population of over 10,000 who were mostly peasants or share-cropping crofters. They grew oats, potatoes and some vegetables. They kept cattle which grazed in the woods (from which they made charcoal to sell to the smelting industry). The cattle were herded on to the moor in the summer, and they were eaten and their hides tanned and used for harness and so on. Evidence of previous cultivation and habitation can be seen all over the island in the form of raised narrow beds which presumably served to help with drainage, and to allow intensive management on the relatively small areas of cultivable land.

During the second half of the nineteenth century the landlords discovered that they could raise sheep on the Scottish hills and that these paid better than the rent from the peasants or 'clansmen' as they were called. Thus many landlords or lairds either sold their land or ousted the peasants by fair or foul means. Sheep were preferred to people. Bayonet, truncheon and fire were used to drive the peasants from their houses. It has been said that the Clearances are now far enough away to be decently forgotten, but the hills are still empty (Prebble, 1963). Many of the peasants ended up in the United States and Canada.

The result of such 'clearances' was that the population fell to less than 1,500 in a period of 20 years or so. In place of the self-sustaining peasant came the sheep, which were able to survive on the hills, albeit at low levels of production. Ten per cent mortality and 50 per cent lambing is still not uncommon in the hills, even with modern knowledge of nutrition and flock management.

However, over the next hundred years the sheep had a profound down-grading effect on the ecology of the area, mainly because they grazed the saplings, so preventing the natural rejuvenation of the woodlands. Previously cultivated areas reverted to grassland

and bracken after the clearances. The practice of burning the heather to encourage young growth for the grouse and sheep in the late winter also aided the spread of bracken and decline of the heather, since the rhizomes (roots) of the former remain unaffected by surface burning.

Since the human population was very low and dependent to a large extent on imports of food and other goods, people were unable to sustain themselves and some islands became almost uninhabited. In others the remaining population was supported by grant aid. After the Second World War, when food production became a major concern of central government, these populations began to be encouraged, not to produce their own food, but to raise more sheep and thus to remain in the Highlands. Over the next three decades the repopulation and revitalisation of the Highlands became an important concern of successive governments. Grant aid reached its peak during the 1970s with grants as high as 90 per cent being given to build houses. The agricultural subsidies have concentrated on increasing headage payments on sheep and cattle. It is still considered by the agricultural establishment difficult or impossible to cultivate these 'less favoured areas' by growing fodder crops, or even vegetables and fruit for human consumption. There are practically no grants available for such activities to bring the land back to arable production. The result of this strategy has been to increase sheep numbers still further.

Ninety per cent of the food for animals and humans is now imported onto the island of Mull, and the only expectation is that it should yield an income from tourism. Much of the land has been sold for afforestation by quick-growing softwoods, such as Norwegian spruce and Japanese larch. Ostensibly these were planned to feed a pulp mill, but in 1985 the mill closed down: it was found cheaper to import pulp from Scandinavia! The several thousands of acres of softwoods on Mull are therefore difficult to sell. At present the main market appears to be as firewood to the local population who used to cut their own peat from the peat bogs.

There are thus strong arguments against the continuation of these types of subsidy in agriculture, which are extremely expensive. It has been suggested that the Highlands and Islands should not be used for agriculture at all, but only for softwoods and tourism. Local populations of farmers would not then have any support unless they became unemployed, which would be cheaper. The National Farmers' Union response to this is simply to put more pressure on government to continue the *status quo* rather than to review the type of agriculture practised and consider an alternative—lower input and sustainable agricultural systems. Farming advice is not thoroughly considered or adapted to local realities. The population grows nothing, or only cash crops, and becomes dependent on food aid from outside the system. There are strong parallels here with some developing country situations.

The Farm at Druimghigha

The name Druimghigha means 'Place of the Geese' in Gaelic. About 50 acres of the farm had been declared a site of special scientific interest (Nature Conservancy Council). Another 50 acres of cultivable land were dotted round the farm in small patches, and at the end of our six-year stay had been brought into the arable/grass rotation. There were another 50 acres of peat bog. The rest was rough moorland grazing with rocks and cliffs. The farm faced north. The varied habitats and topography included river valleys, corries, hanging valleys, cliffs, caves and sheltered woodland. These not only added to the interest and beauty of the farm, but afforded shelter and a plethora of micro-climates.

Some roads had been constructed by our predecessor to allow access to different parts of the farm, and we constructed a few more. Before we left we also divided the farm into blocks by erecting about ten kilometres of fencing over rough ground. Dividing the area was essential in order to manage the grazing and keep animals out or in. The task of creating this infrastructure was not easy! It required considerable resources in time, energy, money and persistence, often in inclement weather (we had two cases of frostbite during the first winter). We built a dairy, forge, horse yard, tack room and living quarters converted from an old shepherd's bothy (a one-roomed cottage), and two conservatories.

Our first task was to provide food for humans and animals. All crops were geared towards this aim, rather than for selling. Small areas were sown with wheat, barley, oats, rye, rape, kale, swedes and potatoes. The cereals were undersown with improved grasses, approximately one-third clovers. The aim was initially to increase the grass growth by manuring. We gradually increased our improved areas. In the end we had ploughed 50 acres (20 hectares), all in little pieces here and there in the hills (Table 15). We wanted to increase the grazing season by early grazing rye, and later grazing rape and kale, depending on the weather and the amount of bracken present (bracken in silage is poisonous to stock). There was also a seasonable abundance of wild foods on the farm, such as nettles, dandelions (*Taraxacum officinale*), wild garlic (*Allium ursinum*), blackberries, wild raspberries, rowan berries (*Sorbus aucuparia*), elder flowers and berries (*Sambucus nigra*). Many fungi, the best of which were the boletus and chanterelles, were prolific in the birch woodland. Birch-sap wine became a speciality.

Table 15. The yearly increase in the area of ploughed and improved ground. This ground occurred mostly in small pockets among rocks, woodland and very steep slopes. There was an estimated 0.2 potentially cultivable hectares remaining in 1989

	1983–84	*1984–85*	*1985–86*	*1986–87*	*1987–88*	*1988–89*	*Total*
Hectares brought into cultivation	2.5	4.16	3.3	3.1	3.5	3.75	20.4

The main industries on the farm were: single, double and multi-suckling beef cattle; dairying (cream, milk, hard and soft cheese, yogurt and butter); poultry for eggs and meat (turkeys, ducks, geese, hens and peacocks); horses (breeding, training, working, competing and selling); sheep for wool and meat (including an indigenous rare breed, the Hebridean sheep); llama rearing for wool, and eventually work. Llama wool is of very high quality and sells well, or can be spun and woven on the farm. Llama appear to thrive well on this montane-type terrain which, except for the high rainfall, is similar to their indigenous Andean habitat. A number of vegetables and soft fruit were grown for consumption and sale, and more than 60 fruit trees were planted, so that overall we had approximately two acres of vegetables and fruit.

The farm was a holder of the Soil Association Organic symbol, although there was little need for this, since the majority of the produce was marketed at the farm gate or in the open market and no extra premiums were charged. Gradually the percentage of our home-produced foods increased (Figure 11).

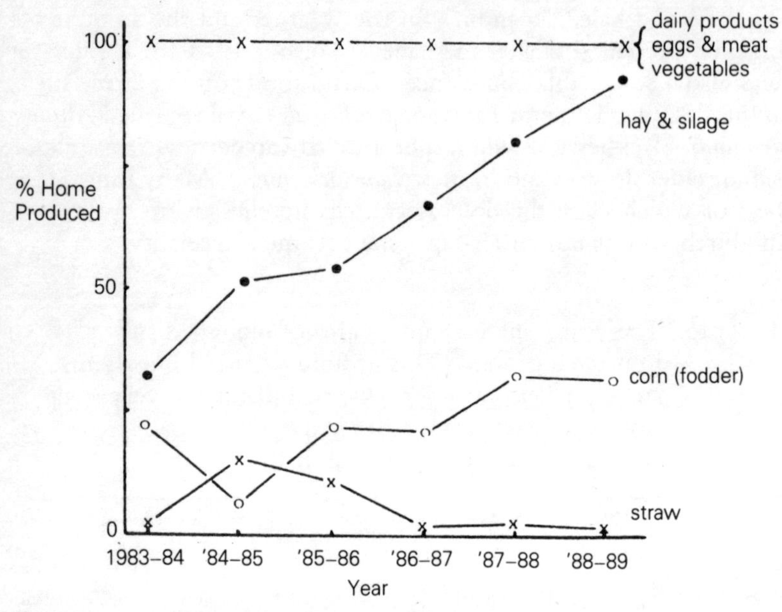

Figure 11. The increase in the percentage of home-produced food and fodder over six years.

THE MAIN PITFALLS AND STRENGTHS OF THE SYSTEM

We worked hard and consistently to make the farm work, but our aim was also to collect information that might be useful in the future.

1 Upgrading the Land

Like many marginal areas, the farm had suffered a net nutritive loss over the previous 150 years as animals had been sold off the system every year and no additional nutrients added. In addition the broad-leaved forests, which maintained and improved the soil, had been removed. Because of the high rainfall and low evaporation rate, leaching of nutrients from the soil had been continuous. Our initial task was therefore to reverse the downgrading of the system (but without excessive capitalisation), and to remain economically viable. There were several ways we could go about this:

a *Draining certain areas*
This would have the effect of eventually increasing pH (degree of acidity/alkalinity). Much of the drainage that had previously been done on the old cultivated areas had become blocked and required renovation.

b *Increasing the soil microflora*
We could achieve this by increasing the humus levels on the potentially improvable arable areas. There was a great shortage of waste humus in the area. Since no cereals were grown there was no straw, and imported straw was costing £75 per ton. We used bracken as a resource in this respect. After cutting and baling it was used as bedding in the winter. It was also used for composting and a small trial area was converted into a rose garden and lawns.

c *Reducing leaching of nutrients*
Increasing the humus levels also helped maintain the nutrients in the soil and thus reduced leaching. Seaweed growing on the shore of the farm was collected and spread, although there was a limited amount we could reach. The sea sand, high in lime and thus useful to correct the low soil pH, was also collected when the tides were sufficiently low.

d *Reafforesting*
We already had over 100 acres (45 hectares) of woodland on the farm. Reafforestation was concentrated on planting shelter belts and on protecting the existing forest from grazing pressure during the summer in particular. Some 1–2 hectares of woodland were coppiced and more indigenous trees (ash, oak, beech), which had become rare, were planted. A hedge was cut and laid along the drive from the old coppice hazel and birch wood.

e *Manuring*
One of the problems with having animals grazing freely throughout the year on large expanses of land of this type is that they deposit their manure randomly and often not where it can be of most benefit. For this reason, as well as to provide shelter, we

built shelters around the farm in which we fed the animals in winter. The muck collected here could then be spread at the time of year when it would have the most benefit. In this way it helped to establish growth on the small pieces of improvable ground, and by applying muck and planting an arable cereal crop undersown with a very mixed clover-rich ley, we managed to grow 80–90 per cent of the animal winter food, and increased the subsequent grass growth by a factor of 300 per cent in some areas, without applying anything from off the system. Thus even in this area, where the growing season is often not much more than five months a year, we could have two grazings and one cut of the grass a year, in the best places.

f *Raising phosphate levels*
Low phosphate levels in the soil are a problem in this area. This is normally corrected by the application of basic slag, or phosphate fertilisers, but we of course did not apply anything from off the system.

g *Improving pasture*
The winters on Mull are long and hard and normally only the number of animals are carried that can be supported through the winter (the sheep are rarely fed at all). This results in over-production during the summer, and hence shading out of palatable species. Increasing summer grazing pressure and the effects of trampling can increase the rate of up-grading (Radcliffe, 1965). Some fields we managed in this way with the indigenous species-rich swards, adding muck and harrowing after grazing. The production increased, but was not as spectacular as after reseeding.

2 **Production of Our Own Feedstuffs**
In the winter of 1983–4 the farm provided only human food. By the winter of 1984–5 it was providing approximately one third of the animal fodder needed and about 70 per cent of the human food. In this year some grass silage was made into small bales and bagged, and also some hay. However, bracken infestation prevented us from making more silage. The vegetable garden had its problems with root fly and carrot fly, but enough vegetables were produced to supply our needs. In the winter of 1985–6 about half of the animals' winter feed was home-produced, and the veg-

etables, meat, fish and dairy products proved more than sufficient for our needs, leaving some for sale. By cultivating rape and rye the grazing season was extended a little, but the season was so bad that we were unable to get the crops in on time, and the silage all had to be raked and picked up by hand! By the winter of 1986–7 three-quarters of the farm's winter fodder was being produced; we had also increased our carrying capacity slightly, and were self-sufficient for human food except for tea, coffee, sugar and oil. During the last two years these figures improved further, so that in the last winter we produced 90 per cent of the fodder needed, although we were carrying more animals.

3 **Self-sufficiency for Labour and Construction**

The number of people at Druimghigha varied between three and 15. All the mechanical, building, electrical, plumbing and cultivating work was done by people on the farm. The only contracting of labour related to road construction and building. The farm supported one full-time worker, one part-time and various volunteers and unskilled students (from 1–16) who helped and were primarily there to learn. Because of the variation in skills required, there were some appropriate opportunities for young and old of both sexes, and even some for the disabled.

4 **Alternative Energy Sources**

Initially we inherited an oil-burning heating system. This was replaced by a wood-burning stove and open fires. We were finally self-sufficient in house and water heating and cooking, although still connected to the national grid for lighting and some power for tools. We also used the glass-house effect to heat some of our living areas and produce food, by constructing two conservatories against the house.

We did not manage to become totally disconnected from the national grid, but about half the agricultural energy used was supplied by horses and people, and the rest from diesel burnt by our tractor. Our energy budgets in 1985 were already much better balanced than those of our neighbours, despite the fodder purchased and our much higher livestock-carrying capacity than usual in this area (Adshead, 1986; Table 16).

Table 16. Druimghigha. Energy budgets and carrying capacity

Energy consumed in the house and on the farm. Because the degree of inaccuracy in attempting to convert different units to standard energy units is so great, they have been left in their original units. The table demonstrates the gradual replacement of non-renewable energy by home produced renewable energy. Relative energy consumption compared to other farms is shown in Adshead (1986). Replacement of the national grid electricity supply by wind generated electricity would further reduce imported energy dramatically

Energy type	*1983–84*	*1984–85*	*1985–86*	*1986–87*	*1987–88*	*1988–89*
Non-renewable, imported						
Diesel fuel for tractor (litres)	1,000	1,000	1,000	1,000	1,000	500
Heating oil for house (litres)	2,700	2,700	0	0	0	0
Electricity (kwhr)	12,000	12,000	12,000	12,000	12,000	12,000
Coal (kg)	0	0	200	200	100	100
Renewable, home-produced						
Wood (kg)	0	5,000	10,000	20,000	20,000	20,000
Working horse hours	520	520	780	1,040	1,560	1,560
Working human hours*	10,400	6,240	6,240	4,160	4,160	4,160

*Working human hours reduced as the infra-structure was built.

Yields of grassland as a result of reseeding or managing indigenous sward.

One of the interesting problems for improving production of marginal area grassland is whether ploughing and reseeding is preferable to improved management practices (topping, spreading manure, careful grazing and harrowing) on the indigenous sward. This table compares the yields for these two alternatives for their total areas over the six year period

Table 16. (*Contd.*)

	Ploughed and reseeded land	*Improved indigenous sward management*
Approximate area	8.3 ha	12.5 ha
Average yield	Cereal silage, underplanted with mixed ley: 6 t/ha	After bracken cutting and FYM (2.4 t/ha): 3.6 t/ha
Yield variation	0–14.4 t/ha	1.68–4.8 t/ha
Treatment (no phosphate or lime)	2.4 t/ha FYM before ploughing	2.4 t/ha FYM in April
Autumn graze	5–10 cm	2.5–5 cm

The yield advantage of ploughing and reseeding is obvious; even the autumn regrowth being much better. However, there was a greater risk and higher costs. In the second year, the reseeded area yields were slightly lower and, by year 3, they were dominated by white clover. A major disadvantage of reseeding over managing indigenous swards was that they were more easily poached. To avoid destruction, all stock had to be removed immediately after the autumn graze.

5 **Conservation, Tourism and Forestry**

The whole farm was managed to maximise species diversity, increase the number and types of habitats and foster preservation with utilisation. The aim was to interweave symbiotically the demands of conservation and food production. Wherever possible, for example, hedges were cut and laid, and indeed some were planted and encouraged, although they were not common in the area. Our long-term policy with the broad-leaved woodland (as mentioned briefly on p. 137) was to coppice it and then to allow it to regrow within a coppicing rotation, in order to crop it again for timber, wood for hurdles, turning and burning. Vanishing or scarce indigenous species of trees and shrubs were planted within the wood, and the woods fenced off to allow for natural regeneration.

The moorland, similarly, was managed as rough grazing with small enclosures to encourage regeneration of species likely to disappear with grazing pressure. This part of the farm, noted for its species-richness, was a carpet of varied colours from May to October. We kept checklists of the fauna and flora, and made a herbarium. A nature trail and farm walk were planned and developed to show the various habitats and tell the walker about the natural and social history, geology and present situation of the farm. There was also farmhouse accommodation serving farm food and running courses for people interested in natural history, harness training, horse-driving, dairying and small-scale and organic gardening, among other things.

Horses were an integral part of the agricultural system, contributing in terms of energy and work. They were used for everything, from gathering the sheep and cattle and working the land to driving to the shop, delivering and competition (including racing and dressage). They gave displays and danced; they also taught and earned money this way. Horses were bred, and young quality horses were sold. Another fundamental reason for having them on the farm was that they were popular. Living in such a rural area, with its particular type of life-style, provided farm-workers with leisure opportunities in which horses played an important role. Visitors also contributed towards farm costs, but they came primarily to work and learn, enjoying the opportunity to sample a type of life that they might otherwise never have encountered or experienced. We had many hundreds of visitors and guests from all over the world while we were at Druimghigha, and it was there that we developed our formal one-year course in the theory and practice of Ecological Agriculture for would-be professionals.

The advantage of Druimghigha from this point of view was that it was primarily a functioning, economically operating farm where the practical tempered the idealistic. The difficulties and trials of living according to what might seem rather idealistic principles could be experienced at first hand. In such a situation primary optimism is a necessity and instils, in some, changes in philosophy and thinking.

6 Economic Viability

Our initial economic predictions were over-optimistic and we obtained neither the grants expected nor the income. The price of purchased foodstuffs was under-estimated, as were the difficulties of the area itself, which added approximately one-third to the cost of all items. We had not bargained for the severity and length of the winter, nor for the unpredictability of the summer. Thus we did not initially keep ahead of our borrowing interest. However, we did in the end manage to pay back the £50,000 loan taken on when the farm started, and to pay off the interest—a considerable achievement with a farm of this size in this area.

The farm survived financially despite the lack of government financial aid; it fed a large number of people and animals, and provided much fun and experience for very many in those six years; but it did not make a major profit (Figures 12 and 13).

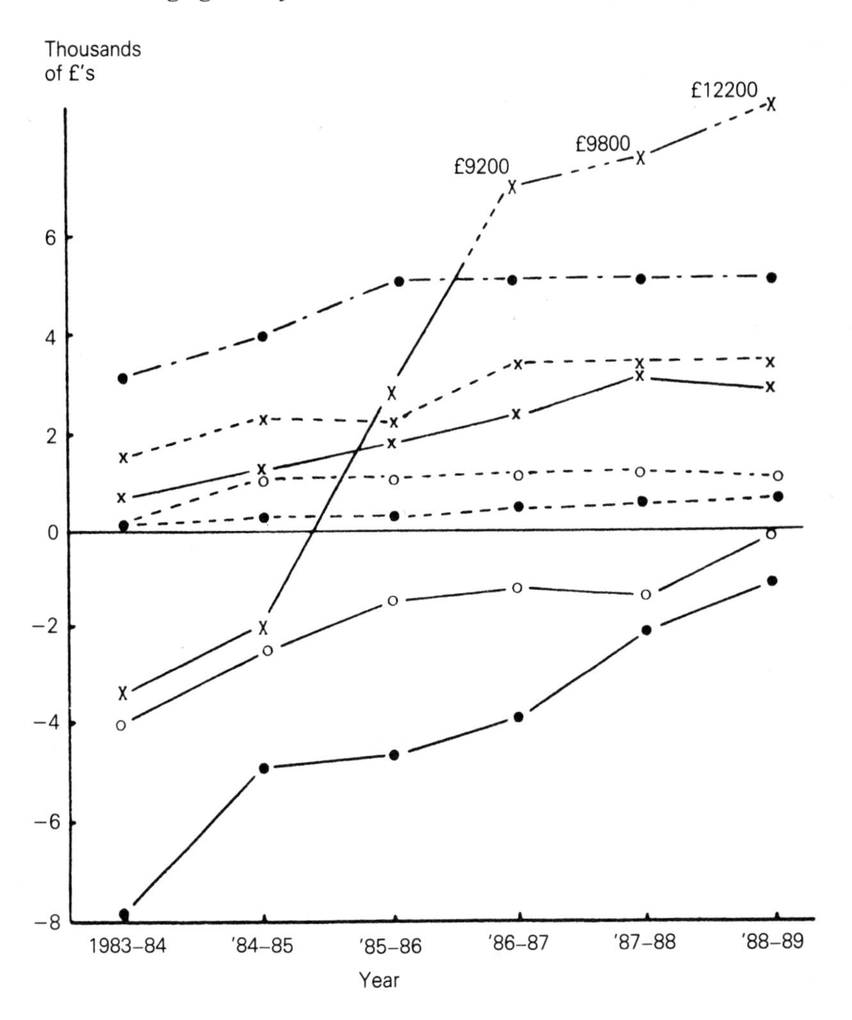

Figure 12. The economic performance of the main farm industries, 1983–89. Above zero – profit, excluding bought-in feed costs. Below zero – total outgoings divided into: loan and rent repayments and bought-in fodder, etc. X —— X, all incomings; ● —— ●, repayment of loan interest; o —— o, corn and fodder bought-in; ● ------ ●, poultry and vegetables; o ------ o, house guests; x —— x, horses; x ------ x, sheep; ● —:— ●, cattle.

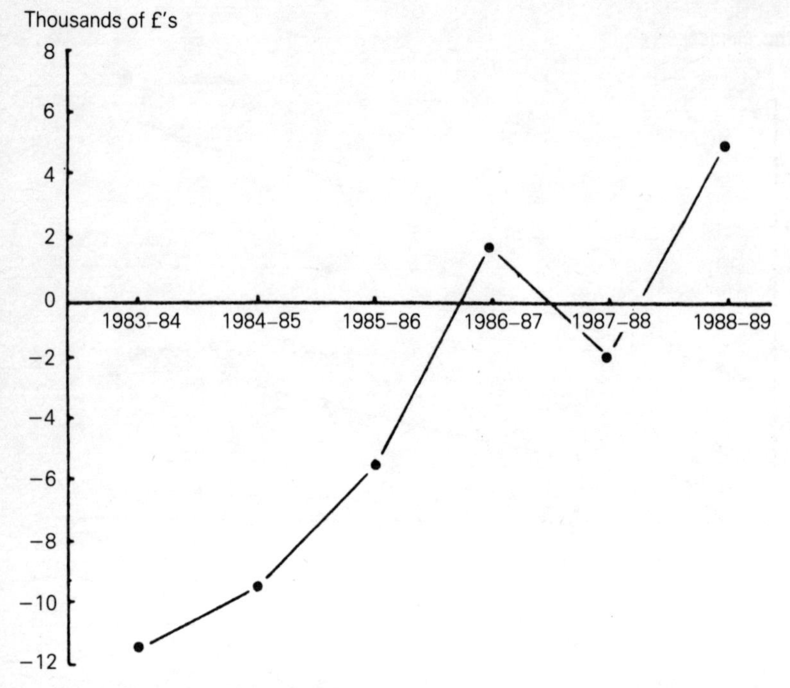

Figure 13. The overall economic performance of the Druimghigha project, 1983–89. The figures include part of labour costs (board and lodging) but not other wages.

7 Animal Ethics

All our animals were kept and managed according to the principles outlined on pp. 86–93. Occasionally some compromises were made—for example, the stallion was not run with the mares all the year as we did not want foals from them all every year; we did castrate some of the young horses and bulls because we could not sell them as entires. We did dehorn the young cattle, again in order to be able to sell them, but many of the heifers we were going to keep retained their horns. The animals had the opportunity to wander over hundreds of acres in the summer; only if it was considered necessary for their good (for example, if they were unwell or in need of extra feeding) were they confined to buildings. Although the weather was difficult, and there was little food around for the animals much of the year, the enormous amount of space in this area made these priorities relatively easy to fulfil.

8 Aesthetics

There is no need for farms to be ugly places, with cheap buildings constructed with no consideration for topography and location, out of non-indigenous materials and with ugly lines. Even the placing of fences, the materials they are made of and the reafforestation, need to be considered in relation to landscape and aesthetic effect. This is particularly the case when the farm is located in a naturally beautiful area. Simple things such as where rubbish and broken machinery are left, where plastic sacks and baler twine, old bottles and barbed wire are dumped are often aesthetically disturbing.

Nowhere is this more the case than in areas such as the Highlands, Australia and parts of the United States, where there is apparently much space and little pressure of population. The result is often a frightful rural squalor. The impression of the neglected 'down and out' farm is conveyed only too often—even when the farmer drives around in a Mercedes Benz! Although the farmer and rural dweller may not often notice this, it does tend to affect him and his behaviour (see chapter 11, p. 239). Self-sustaining farmers, on the other hand, can often recycle or utilise 'rubbish'. The challenge is to make the rubbish pile look ordered and part of the landscape, or well hidden.

* * *

The main strengths of the ecological farming system for the Highlands are that the area is potentially very productive and biologically rich, despite the gloomy prognosticatians of agricultural advisers and the conventional agricultural establishment. However, careful strategies are necessary to upgrade the system using only its own resources. The use or creation of micro-climates is crucial, and planting times are also very important. A large diversity of habitats can be made use of and can be integrated with conservation. The area could support a much higher population and provide employment without the need for large amounts of government financial aid. There is a ready market for all the farm products to feed local people, and to cater for tourism if necessary.

The difficulties of the system include little or no infrastructure for this type of farming, so that equipment is largely unavailable for making buildings, fences, farm roads and tracks. This adds very considerably to the costs and difficulties of operation. There is a lack of organic materials that can be used as the building blocks for upgrading the system, but again, this problem can be circumvented, although the upgrading may be slow. There is considerable pressure from wild animals grazing. Fences must be strong and elaborate and so are expensive. The main trouble comes from marauding and athletic hill sheep, rabbits and red deer who are able to hop over normal fences. Crops destroyed by these animals in early plantings, either on a garden or a farm scale, are one of the major disincentives to development in the area. There is also a gross lack of information on varieties that might do well there, on planting times of horticultural and other crops, recycling of nutrients using only the resources of the farm to prevent leaching, and so on. Conventional agricultural advisers and research establishments are of little help since they do not at present have answers to many of the questions raised by the ecological farming system.

Over the years of developing and running Druimghigha, certain points emerged as of great importance in trying to help oneself or others in difficult marginal areas of the world. For a start, the area could be made very much more biologically productive from its own resources. The controls are not fundamentally biological, the 'poor soils' and 'bad climate', so often quoted as the reason for dependence on government support, are not crucial, although appropriate environmental management is.

The main stumbling block, and the reason why many others have not developed farming along ecological lines in the Highlands, is largely cultural. I shall never forget the extreme negativism of, in particular, the advisers who can have, and indeed have had, an enormous effect in demotivating the population from trying anything remotely new or different. In a culture which remains strictly hierarchical and inclined to wear hair shirts, it is crucial to understand the importance of constructive encouragement.

What seems central to the success of such projects is the attitude of those doing them. Negative thinking never built the Taj Mahal. Positive thinking, on the other hand, can turn mountains to molehills, combined with whatever relevant scientific knowledge and experience available. But be cautious here! I was told by scientists in the most relevant research organisation in Edinburgh that we would not be able to establish clover and improve our hill pastures unless we put on at least lime and basic slag or phosphate in some form. This proved to be incorrect. Two years later the clover covered 80 per cent of our well-grown sward and, nine years on, 60 per cent. Had these scientists not done the research, or had they done it badly? Like so many belief systems in science, it had, it seems, just gone down in the literature; thereafter it was assumed to be true, and no one had tried it since someone had failed long ago!

Druimghigha had fulfilled many of the tenets of Ecological Agriculture, rather more than I had believed it would. It was time to move on again.

7 The Next Phase: Little Ash Eco-Farm

We went to Druimghigha to try out our ideas in practice in a challenging environment which, like many other marginal areas of the world, was in dire need of solutions to agricultural problems. We gave the farm six years, and I think what we achieved was more than interesting, it was remarkable—not because we were particularly good farmers, but because it emphasised how benign the living world can be, if one tries to work *with it* instead of imposing on it what might be nicknamed 'Anglo-Saxon attitudes'.

For me, biologically, philosophically and culturally, it was a great environment in which to learn. After six years, however, we had done the basic hard work, the farm was functioning well and we felt we had come to the end of the experiment. By 1989 economic crises and public opinion were combining to make even government reconsider its traditional support of agriculture in the Highlands and Islands. We had shown that there was a workable alternative, provided farming and farming attitudes changed. It would help solve the area's economic, social, environmental and conservational problems in the future, while also increasing employment and rural-based living. Perhaps it would even ensure that more people owned, or had some control over, one of the crucial resources . . . land.

We had contributed our two pennyworth: it was now for public money, in the shape of government-financed research, to continue and develop this. We wished to move to pastures and ideas new. There were many reasons for this, not least the fact that although the humans, cattle and horses had adapted well to this environment, life for us all was tough. None of us liked the predominantly cold, wet weather very much; the winter was a tough, long and dark time in those latitudes. As one who had been raised in Africa, with a different climate and different cultural attitudes, I found it a difficult place to live in. Besides, we had shown that Ecological Agriculture could work, even in marginal, difficult areas. We

Baksheesh, Barney and Druimghigha, Mull, with the mountains of Rhum in the background—what Eco-Agriculture is all about.

The semi-pastoralist life—where to now?

Camels in Natu, Kenya, with pastoralist Turkhana.

Aesthetic problems on the Downs—an ugly building, badly sited, on a farm professing to be organic, judging by the abandoned fertiliser bag.

Mechanical weed control.

Clover sward on improved soil at Druimghigha, Mull.

Milton Court Farm: the nature reserve 18 years after it was first established—now named an English Heritage site. Note the young yew sapling in the foreground.

The old farmyard at Milton Court.

The forge, built and operated by Tony C.

Ballo the cow being milked by Kathy G.

The cattle shed and yard at Milton Court, which we rebuilt in the traditional style.

Horses housed appropriately—in family groups in barns and yards—not isolated in single stables.

Post-and-rail fencing made from coppiced chestnut trees makes a sound barrier behind which a hedge can be grown to replace it eventually.

Druimghigha: the varied topography and vegetation.
Photo: J. Goldblatt.

Druimghigha farmhouse.

Shifting muck is a job in which everyone helps.

Cereal crops flourishing on the upgraded soil.

The homemade forge we built at Druimghigha, from our own timber, recycled tin, local stone and an old vacuum cleaner. *Photo: J. Goldblatt*.

Double suckler with horns.

Prototype multispecies living, Little Druimghigha. The windmill provided electricity and water was drained from the gutters into recycled plastic tanks.

The ram pump we installed at Little Ash on Dartmoor—an ingenious West Country invention.

Little Ash: the multispecies dwelling in operation.

Wild flowers recolonising at Little Ash.

Milking sheep. *Photo: John Lyne, Tamar News*.

Shearing llamas.

Animal ethics 1: sow stalls in an intensive pig unit.

Animal ethics 2: pigs living in a family group. The sty is on the right and the pigs have a spacious run and are let out in the fields some of the time. *Photo: J. Goldblatt*.

Kikuyu women who came to dance at the author's wedding.

Japan: cattle housed in small sheds.

Village near Myasaki. Here a basketmaker works with local bamboo . . .

. . . and free range hens scratch about under the mulberry trees grown for silkworms.

Saline soil in the wheatlands of Western Australia.

Desertification in Australia.

The remarkable regenerative properties of Australian flora are evident in the new growth on this felled gum tree.

wanted to progress to developing the idea further . . . starting again in another place.

In the last decade of the twentieth century, one of the major problems confronting agriculture worldwide is how, if at all, it can integrate with wildlife conservation and in particular National Parks. Is there any way in which humans can live more symbiotically with living systems, even indigenous ones? Can they integrate food production with nature conservation?

There is a rising tide of 'nature conservation separatists' (N.C. Apartheid for short!). These are people who believe that it is vital to preserve the 'natural' environment. The wild fauna and flora must be left undisturbed by human beings. The reasons given for this approach are: that future generations will need to see what the wild-world was like; that it will represent a gene pool which can be drawn on to solve human problems in the future; that it is not possible for humans to live in association with the natural ecosystem since they inevitably destroy it; that nature is too special and alien—it must be admired and worshipped by humans, preferably from a distance, or if close at hand only so long as the humans do not interfere in any way.

Yet at the same time as upholding these arguments for the creation of nature reserves, the voting, paying public consider that they should have access to them. There are two major problems with this:

1 Even though they walk through them only to admire and wonder, perhaps to eat a picnic, they are going to affect the environment, whether or not they pick plants, ride their horses, row their boats or sleep in them at night. Why, if they really believe their own arguments, should they visit at all? Surely a few TV films would be enough?

   Even if it is argued that a close experience is necessary, where is the line to be drawn between what is permitted and what is not? Is it necessarily the case that rowing a boat or riding a horse through the National Park will cause more ecological damage than walking and picnicking? The rules and regulations in different National Parks throughout the world are often not questioned or even considered.

2 By isolating areas in this way, ecological changes will inevitably occur because, unlike a work of human art, natural biological systems are characterised by change and evolution. For

example, the Downs in the South of England are well-drained chalk hills whose climax vegetation type, 2,000 years ago, was largely beech-dominated mixed deciduous woodland. The clearing of the land, and the grazing of the resulting grassland by sheep over 1,000 years or so, has resulted in a particularly species-rich 'downland turf', including orchids and many other relatively rare, delightful chalk-loving flowering plants. When areas of the downland were made into National Parks, or 'sites of special scientific interest' (SSSIs), the sheep were removed since, after all, they were domestic animals and not the indigenous fauna. The result: the downland turf was lost, overtaken by a host of bushes and creepers—a step on the ecological succession back towards the deciduous woodland.

What should the National Park authorities do? Cut the grass with mowers? (They tried this: it was expensive and did not have the same effect as the grazing sheep.) Consider this a natural ecological development? (They tried this, and the rate-payers shouted loud and clear that this was *not* what the National Park had been created for!) Finally they put the sheep back, thus admitting that, in this case, the reason for the National Park was not its natural ecology, but its *man-made* ecology . . . and why not? But we must realise this is not in line with the purists' belief and understanding of what Nature and National Parks are.

Even in the few parts of the world which have not yet had their ecology changed dramatically or destroyed by human beings, such as a few small areas in Australia (the population of the whole continent is 17 million humans), human management is often crucial to the maintenance of the reason for the National Park. For example, fire in the Australian bush, either naturally occurring or, in the last few thousand years, the result of Aborigines' management, is necessary for many species to flower, or seed to germinate. If the area is shut away as a 'National Park' and fire prevented from spreading to it, then the ecology changes dramatically over a period of years . . . This is, after all, 'natural', but again *not* in line with what the National Park was created for: to preserve this particular patch as it is . . .

★

These problems are not idle silliness: they are at the centre of the public and the National Parks authorities' confusion. There are many problems and debates that must be confronted, but these particular attitudes rest, it seems to me, on three specific illusions:

1. That humans are not really part of nature and cannot and should not be . . . despite the fact that we are mammals, eat, drink, defecate and copulate like others.
2. That it goes without saying that humans cannot be integrated with the natural ecosystem, since they must inevitably destroy it.
3. That it is possible to 'freeze' biological ecosystems in time. This, despite the fact that *by their nature* biological systems are dynamic, changing and evolving.

I am not convinced that we need to embark more fully on the path towards apartheid from nature. There are those who would prefer this, and because individual liberty is important, they should be able to practise it. Does this mean, however, that they should impose their values on us all? It seems to me that we should be able to integrate human needs with self-sustaining natural ecosystems, although this may mean changing or modifying various cultural priorities and our profligate use of resources.

In some parts of the world National Parks are created where the indigenous human population has either departed voluntarily or has never lived. In others, the indigenous population is kicked out in order to create the National Parks. In Britain (population 56.2 million humans in an area the size of one state of Australia) there are people, and have been for the last 2,000 years, who own and/or live all over this densely populated country. Since there is universal suffrage, neither missionaries nor dictatorial governments can get away with throwing people off the land to make National Parks. The result is that somehow or other National Parks have to come to terms with the people living within them. In some cases one of the aims of the National Park is to preserve the social and aesthetic characteristics of the area by encouraging a rural life-style, land use and architecture of the human residents as well as the others. One of the main stumbling blocks with this approach has been how to do it. Most of the land is farmed, and the rural-dwelling farmers have been trained to farm according to the current modern high-input farm industry model. This *by its nature* does not take account of the natural ecosystem but is committed to environmental manipulation to maximise profit.

Problems are inevitable, and there will be clashes over priorities and planning within the National Parks. How can these be resolved? Is there a blue-print that can be used? It may be that Ecological Agriculture can help to integrate land use and conservation, and clarify the situation for National Parks in many parts of the world.

We had worked out and run the animal management on our farms to fulfil our own rather radical criteria, taking into account the ecological, behavioural and ethical needs of the environment and animals. Better still, we had shown that this also worked economically. But the longer we did this, the more unsatisfactory we found it. Gradually my partner and I evolved our thinking towards a system that would give animals and humans more equal consideration. Why should humans inhabit relatively spacious houses, with several rooms full of furniture and ornaments, segregated from their animals, fed with the national grid and water supply, dependent on the county waste disposal system, built and furnished with materials from all over the world whose production often caused environmental problems locally and globally? Meanwhile their animals were crammed into large, ugly barns, often without bedding to live on, grass to eat, or a chance to wander and mingle as they chose. In any case, did these animals inevitably prefer each other's company, or would they, given a chance, sometimes choose to associate with humans, and delight in learning and experiencing new things?

Were we able to design a mutual living system in which we could all share more and be self-sustaining not only in food and fodder for the animals, but also in power, water, waste and materials, and with the humans in particular being less profligate with living space and possessions? In this way, could we learn more about the animals and how they perceived the world, and teach them about our ways? Would we all thereby learn more and enjoy each other's company . . . live more symbiotically? If we could design this system within a National Park, would it work?

At the end of the day, one's purpose in living with animals on a farm is often that they will be killed, sold and eaten—although one hopes that they will have had a happy life, relatively free of suffering and full of joy and experiences. The more I was confronted with the reality of this in relation to the small number of animals we had on the farms, the less acceptable this became to me. Were there other ways in which we might be able to make a living with our animals, without having to have so many killed? There were certainly other ways in which different animals and species could contribute to the farm income by, for example, milk and milk products, eggs, energy and power, enjoyment and recreation and, of course, fibre. Had we really exploited all these avenues sufficiently for all the species we kept? Finally, if we did have to raise some animals to be killed, would it not be possible to raise such high-quality, able and delightful animals that they would be bought at high prices for breeding rather than for slaughter?

These were the questions that now confronted us, and our next project must obviously be to try out some new ideas. The philosophy and practice of food first Ecological Agriculture had proved successful and relevant to many different countries and places . . . but could it evolve further?

In December 1988 we sold the greater part of Druimghigha, keeping 70 acres next to the sea to which we moved in order to build the prototype of our multi-species dwelling. The next three months, with students from Poland, England and Scotland, were spent starting afresh. Our highland multi-species croft must have its own water supply (rain water saved), its own power (a small windmill), provide its own fuel (wood from the birch and hazel woodland), recycle its wastes, grow its own food and create an environment in which, even during the west coast winter, humans and animals could live happily together, warm and well fed, comfortable and with mental delight and joy.

Washing, other than in the icy sea (which was not conducive to frequent repetition), proved difficult. Heating a kettle to provide a small amount of warm water was no hardship, but then followed the draughty, rapidly splashing wash—not pleasant enough to repeat too often!

One day we had the idea of constructing a bathroom with technologies from my Kenya childhood. It took shape, an extension to the rear of the multi-species residence, with a 40-gallon drum on its side fed by the roof water and heated underneath by a wood fire lit in a specially made hearth. There is nothing like 40 gallons of boiling hot, rather brown water bubbling into an old metal bath tub someone had given us for a drinking trough, with billowing wood smoke filling the well-windowed and naturally ventilated amateur dry stone-walled bathroom. To lie cooking in that bath neck deep, flesh red, muscles aching, looking out at the wild Mulluch winter rain and wind after a hard day of building, planting, carting, chopping, talking and milking, beat any other bath I have ever had.

By February we had built and tested our prototype. There had been uncomfortable days, and problems with design and building, but it had more or less worked: we had devised a pleasant, comfortable arrangement whereby we could all keep dry and warm, eat, laugh, shout, read and learn. It was time to move south, to put into practice our ideas of many a year, and design and build our new multi-species farm.

Moving the farm south was no easier than moving it north. Let me caution every reader: never move a farm if you can avoid it! There are not only the normal human household effects, but farm machinery, irreplaceable bits and pieces, the household effects for horses, cattle, sheep, llamas, dogs, ducks, hens and everything else you might be moving—as well as all the various tools, food crushers and grinders, lifters, spreaders, mowers, welders and potentially useful objects, including an array of empty oil cans and old metal buoys (just the thing for making wood-burning stoves). We hired an articulated lorry which arrived late on a snowy night off the ferry.

In the dark, icy predawn we loaded the cattle, their calves, the sheep and finally the horses. I had been debonair about the lorry driver's questions about whether the 18 horses would be any trouble to load onto a five-foot-high artic. in the dark ice and snow. Good heavens, I ran a consultancy on behavioural problems of horses, mine at least should load! Oh, how the mighty fall! All four of the chestnuts took extreme umbrage and refused to enter the lorry. Time was passing, the lorry had to leave to catch the 8 a.m. ferry or we would have to pay another £1,000, and still Shiraz, Karma, Sheira and Aroha refused to go in, despite their relatives all assuring them of the acceptability of their accommodation. It was touch and go, but finally, by moving the lorry (I knew it was not quite right but had not expected Shiraz to be such a pedant), they were in and the massive vehicle was cautiously sliding down the icy glens on a road six inches narrower than its own wheels, followed by our own lorry full of the poultry, llamas, dogs, furniture, tractor implements and so on, towing a caravan.

Behind us came the students, driving our overloaded car pulling another caravan . . . We caught the ferry, but then on the mainland the caravan blew up, and we left it in a highland glen to collect another time, and rushed on through the night to try and meet up with the artic. and the animals. All the driver had was a map drawn on the back of an envelope, on which we had indicated the field in Devon where he was to unload the animals. At 3 a.m. we arrived to find that he had been and gone. They were all in their new home, munching knee-high green grass such as they had not seen for five months, smiling all over their bodies.

It rained for six weeks and the building had not begun, but even then, compared with the Mull winter it seemed to us like the land of milk and honey. We had a contractor with crane to put up the main frame and roof. When the roof was on we thought we would be in clover—a 50 × 100 ft dry space! Not so lucky: the levels were wrong and all the water drained inside, making a lovely muddy lake for our son to play ships and navy, but somehow missing the point of the roof!

Hiccups, hard work, disasters and successes continued over the next two years as we gradually built our farm, made mistakes, rectified them and made more. Our designs, drawn on another envelope, gradually altered and evolved as we progressed. But for the first time we were not inheriting someone else's ideas and design, however old and historic; we were designing our own for the twenty-first century and trying to fulfil still more closely the requirements of our own tenets. The materials must be supplied if possible from the farm itself. If not, then their manufacture and supply must not cause global or local environmental problems. We knew that we could easily run the farm in a self-sustaining way, but could we be self-sustaining in energy and water?

Water

Despite the relatively high rainfall and low evaporation rate in Britain, the last few years of lower than average rainfall have pointed out the vulnerability of our present water system: the profligacy of water use by humans. Rationing, and banning of sprinklers and hoses, were introduced in many areas. Yet still more valleys are flooded to store yet more water for urban humans, amid protests from rural dwellers and conservationists. Even now the system is not being seriously examined. Why do we continue to have waterborne sewage? With the technological knowhow on which we pride ourselves, are there not other ways of dealing with it? Washing machines and dish-washing machines are extremely wasteful with water. Even with our relatively easy system with a standard washing machine, we can only do one load a night. Surely there are ways of improving the design and reducing the water and electricity consumption, but still allowing us to have them?

Water was our first and most important requirement. We found an old ram pump in the wood, and reconstructed and re-routed pipes so that, with no energy input but that of the falling water itself, ten per cent of the water was pumped several hundred feet up and stored a quarter of a mile away in recycled orange juice plastic tanks planted around with trees to hide them. From there it fell by gravity to all the troughs throughout the farm, and to the building. Ram pumps, invented in the West Country, are amazing things, simple but ingenious. They can be built by almost any peasant and can save millions of hours carting water . . . an imperative technology for much of the world to rediscover. We have had teething problems with our ram pump but now, three years on, it is working famously and with luck we shall even have water for irrigating next year. We could also save the rain water. If we were to do this, we would have enough for all our own and our animals' domestic needs during the year. We collect some of it to show how possible this is, and use it to feed our only waterborne lavatory—a necessary addition if we have visiting parties of schoolchildren, as we were asked to have within the first few months of our arrival.

Electricity

Lights are possible without electricity, but there are some machines that make life a lot easier, for which it is essential. In a temperate climate, where one has to wear many clothes much of the time and washing does not dry overnight, a washing machine and even a drying machine make an enormous difference. Crushing of animal food and grinding of wheat for flour can also be done by hand, but after several years of doing this from time to time, why not use appropriate technology and have an electric machine so that one can do something else instead of grinding the flour? A welding machine to mend the constantly breaking bits of metal for farm machinery and buildings is also irreplaceable. And what about television? Why should one not enjoy a good film or an irritating political debate from time to time? Finally, what about my computer on which this book was written? Is it not possible to have a PC powered and manufactured acceptably?

For us, in our attempt to develop a 'purist' ecological farm, the national grid electricity was quite unacceptable. Firstly, it uses non-renewable resources such as coal and gas; secondly, a percentage of it uses nuclear fusion, a recipe for environmental disaster; thirdly, the central location of the generators demands that pylons stride over the countryside, carrying electric cables in all directions with little regard for the visual aesthetics.

We did not have a river flowing with sufficient energy for hydroelectricity, but on Dartmoor, in the National Park, we certainly seemed to have spectacular winds, and especially in the winter when we needed most of the power. As a stop-gap we rigged up a small windmill on our barn roof to charge a 12-volt battery, and with a small inverter it was not long before I could use my computer. It also allowed us to continue working after dark every now and then, with the aid of a 12-volt light. We thought it would be only for a short time, as our intention was to have a four-kilowatt windmill, suitably placed to avoid making an environmental scar. How wrong we were. We had thought we would have to put our case to the planning authorities of the National Park, but we did not know how difficult and powerful some of the local opposition could be!

After two years of struggle, we now have our legally working one-kilowatt windmill and accommodation. Provided the wind blows, we can run most of our essential machines on the electricity, if we are careful. Most of the manufactured machines (washing, drying, washing-up, fridge and freezers) use a large amount of electricity. Is it not possible to design them to use less and still be efficient? For example, we tried, with no success at all, to obtain a deep freeze that was sufficiently insulated to remain frozen for 24 hours without electricity. We have failed and had to give up freezers. (See Table 17.)

Building Materials

When it came to materials for building, we acquired a marvellous selection from a mental hospital which was being torn down to be replaced by modern houses built largely with imported

Table 17. Energy budgets for Little Ash, 1992 (figures worked out from Blaxter, 1975; Kiley-Worthington, 1984)

Animals and Humans maintained (fed entirely off the farm)
11 cows, 18 calves and bull, 15 store cattle
88 ewes and rams
20 poultry
10 horses
3 dogs
5 people (average through year)
10 llamas
5 rabbits
Total: 310,347 MJoules

Animals, animal products, plants and plant products produced

Sold off farm	*Consumed on farm*
15 finished cattle	
100 finished lambs	2 lambs
1 horse	
	625 eggs
	20 poultry
1,300 l milk products	1,300 l milk and products
98 kg various wools	
1 ton vegetables and fruit	2 t vegetables and fruit
5 ton cereals	5 ton cereals
	100 t silage
	50 t straw
Total: 251,700 MJ	27,397 MJ

Total produced: 279,097 MJ

Energy inputs to farm from off it
Diesel 3,600 l
Transport 200 ton miles
Total: 154,128 MJ

Energy produced on the farm
Wood 2 tons
Wind 2,880 kw
Horses working 13,500 MJ
Humans working 4,500 MJ
Total 82,200 MJ

Total energy used on farm
154,128 + 82,200 = 236,328

Total energy used produced on farm minus that bought in
154,128 − 82,200 = 71,928 MJ

Total energy produced by farm, excluding that consumed
251,700 − 71,928 = 179,772 MJ

Table 17. (*Contd.*)

Energy produced per hectare excluding that required for maintenance
179,772 ÷ 33.3 = 5,398.5 MJ

Total energy produced and maintained on whole farm
310,347 + 154,128 + 82,200 − 71,928 = 474,747

Energy produced and maintained per hectare
474,747 ÷ 33.3 = 14,256.6 MJ

For every bought in calorie from off the farm, 2.49 cal are produced to be sold or consumed (179,772 ÷ 71,928)

While 14,256.6 MJ are produced and maintained per hectare (excluding standing plants and wild animals)

materials and powered entirely by electricity. We bought second-hand floorboards (it took one week to remove all the nails), plenty of glazed windows that were being thrown out and destroyed—no market, you see—and many other useful materials. My father gave us some tree trunks blown down in the 1987 gales in Sussex. These were oak, ash and beech timber trees. They were not of sufficient quality to sell as furniture, so, like thousands of tons of beautiful hardwoods felled in those tornadoes, were going to be cut up and burnt! We had them hauled out of the wood and cut up into planks, collected them in our small truck and set about building modern oak beams into our harness room, dairy, granary and feed store. The idea was to use recycled or otherwise unwanted waste materials for our buildings, but we had had no idea we would have the opportunity of lining the human part of our multi-species abode with ash, building oak shelves and beech window boxes! It made aesthetic and economic sense in the end, too.

Landscape Design and Conservation

One of the tenets of both our other farms had been to think seriously about environmental design and to ensure that the whole farm was a conservation area—to try not to eliminate any other species, but rather to work out a way of living together so that we could all manage. Thus, coming to a National Park where such a concern should be central was no new experience for us. We first set about reducing the sizes of the fields which had had their hedges removed in order to make the large dairy farm easier to run and more profitable. We obtained old maps of the area, and planted hedges where they had been before, with accompanying fences to avoid the young plants being eaten. To avoid the fences being a dominant feature of the landscape, and to save on materials such as wire netting, we have used wind-driven electric fences wherever we can; by the time the hedges have matured, these will have more or less disintegrated. We fenced off the woodland and planted more indigenous trees. Over the years much of the species-poor secondary birth and alder woodland will be coppiced, and the wood given a helping hand towards achieving its climax by planting more ash, oak, some beech, and various other indigenous trees.

Of crucial importance have been the placing and attempts at visual integration of the large building. The reason why we chose to have such a large structure was because there are very considerable advantages in the ease of running a farm if it is all under one roof. Our multi-species system demanded that we should also try to fit the human dwelling into the building, as well as all the farm offices.

Because of their convenience, large farm buildings are preferred by most farmers today. The problem, therefore, was how to integrate a large farm building into the landscape and make it aesthetically inoffensive if not aesthetically pleasing. We have done away with any service cables or pipes (telephone underground, home-produced electricity, underground from windmill to farm). The only cables we have on the farm are those running across one of our fields, installed by the electricity company before our arrival to supply a neighbour. We shall fight to have these placed underground in the next few years.

We discussed in depth the placing of the building with the National Parks authority before we finally decided to site it behind an existing well-grown hedge in a slight decline, and plant tree screenings around. On the building itself we have planted quick-growing, evergreen, highly scented, hardy and beautiful flowering creepers, and we shall eventually cover the cement sheeting around the skylights with turf to create a wild hay meadow on the roof. All these extra costs, including extra excavation, roadway and drainage, have been considerable, but the result is that the building can hardly be seen from the road or anywhere on the farm . . . even from the air it is hard to spot in comparison with some neighbours' large modern animal housing systems. Thus, hardly surprisingly, it seems that it is possible to have large, convenient farm buildings even within National Parks without causing environmental or aesthetic degeneration, provided their siting, construction and screening are carefully thought out and implemented.

We inherited a well run but intensively managed dairy and chicken farm. The grassland had been managed to encourage the few quick-growing, milk-producing grasses and discourage indigenous herbs and grasses which were not particularly palatable or good for milk production. So-called 'weeds' had been sprayed with herbicides, the grass had been treated with up to eight hundredweight of nitrogen phosphate and potassium fertilisers a year, and slurry had been pumped out of the cattle cubicle housing and the intensive chicken houses at the rate of tonnes a year. The result was that most of the wild flowers and indigenous flora were absent in the fields, and there were no pansies, campions or cow parsleys in the hedgerows. In the first spring we scattered seeds from many of these common herbs in the hedgerows, and shaws, and surprisingly quickly they are galloping back—red campion, rose bay willowherb, thistles and dock are having a field day. Most importantly, the clovers have almost taken over in the meadows. Other useful herbs are being planted in the swards as they are ploughed up to enter into the rotation (once in every eight years).

The variety of wild birds and mammals is increasing, too; we do not have any shooting or hunting on the farm, and with the increased variety of plants around the barn, many of the garden song birds are taking up residence. In the barn the robins and yellowhammers vie for territory and shout and sing their success to the world in the spring, descending in droves to eat up the spilt food of those large and clumsy friends, our bovine occupants. The squirrels are gradually gaining confidence to steal the remains of the dogs' food from the lawn next to the wood. Duckie Pool wood has always had a considerable population of badgers, now joined in plenty by foxes. Fox and badger city undermines the woodside of one of the fields, adding to the excitement of mowing the field—will the tractor drop into a fox or badger tunnel, or continue unscathed to the end of the row of the wood edge? The foxes have had a few Sunday lunches off our ducks and poultry and it is tempting to try and get rid of the wretches, but on the other hand, should there not be enough for all of us, and should we not ensure that the poultry are shut up at night and cannot be gobbled up? Sometimes it's hard to handle, but of course we should indeed do this . . . perhaps we'll manage it more efficiently in the future.

Multi-species Accommodation

So how does this work? The theory is that during the winter, when they are in the barn, each species should have its own area and be able to go out of it at will. The animals should have access *ad libitum* to the appropriate feed for themselves, and they should be in the appropriate group sizes and structure. They should be allowed to mother their own offspring, and have sex as and when they wish, and they should be able to perform all the behaviour in their repertoire which does not cause suffering to others (chapters 8 and 9).

In addition to these parameters, which we were able to fulfil at Milton Court and Druimghigha with our animal management, given that these animals have a life which is more or less in line with the type of life they have evolved to live in (although it may be rather more comfortable since they are never as hungry, thirsty and cold as they might be in the wild, and if they are sick they are given drugs and treatment to try and relieve pain and suffering), we are asking if they are content always to choose their *own* company rather than that of another species. Are there individuals who will sometimes choose to join other species, perhaps even choose to have rather more intellectual and physical stimulus in their lives?

We are also asking questions concerning the cognitive (thinking) abilities of the various species. We know they think; the problem is, *what* do they think and what can they learn in the human area of intellectual abilities? Similarly, what can we humans learn from them? After all, the point of our life here is to live symbiotically with each other, to our mutual advantage, and to learn from each other more about the world we live in. The experiment has only been going a few years so far, and we have been building and organising the farm most of the time, so we do not yet have answers to these questions.

What about confining humans to the same sort of space per individual that our animals have? Do we really need houses with several rooms full of possessions of various sorts, or is it possible to cut down our space and resource uses dramatically without a serious decline in our happiness, satisfaction, health, length of life, and culture? In other words, are monetary income and accumulation of material wealth and possessions satisfactory criteria for measuring human success and happiness? This must be true for some, but if one is fundamentally interested in the living world, in integrating one's life-style appropriately, and in particular in living with other sentient beings as well as humans, then there is no conflict. I am delighted not to have a house full of objects of worth and value which have to be worried about, cleaned, watched and insured. I don't need the works of humans around me; I have much more aesthetically mind-blowing experiences every day . . . a blue tit, blue feathers catching the early morning sunlight, hops onto the lawn, and the dawn has a hundred colours constantly changing . . . a foxglove's flowers of pink poetry carry me away in a dream . . . The magic of the living world has the added aesthetic value of its dynamism and impermanence—it is constantly changing, a brief experience to be plucked and remembered; an experience of a bigger and brighter jewel than ever a magpie or a jeweller owned darts shimmering from one tree to another, tail waving in the evening crystal, black eyes watching, moving, enthusing and telling deep tales . . . as great as if not greater than those of the Mona Lisa. Equine arab faces nod gracefully as their owners glide across the greener field than ever an artist could convey, their grey and golden coats shimmering . . . a moment caught in mindsummer madness.

No, there is no problem with living closer to the living world, in fact there are many compensations. Perhaps other humans from the twenty-first century might try it too. It is not necessary to be 'wild, uncivilised, uncultured, non-intellectual primitive people'. Sometimes I wonder who is primitive, and what, oh what, is progress?

The Farm

The farm at Little Ash presents no problems. Relatively it is a land of milk and honey, although it is supposed to be a 'less favoured area'. Within our first year we were producing all our own and our animals' food, and some to sell.

We have introduced several innovations to try to develop a way of making a living out of our animals without selling them for

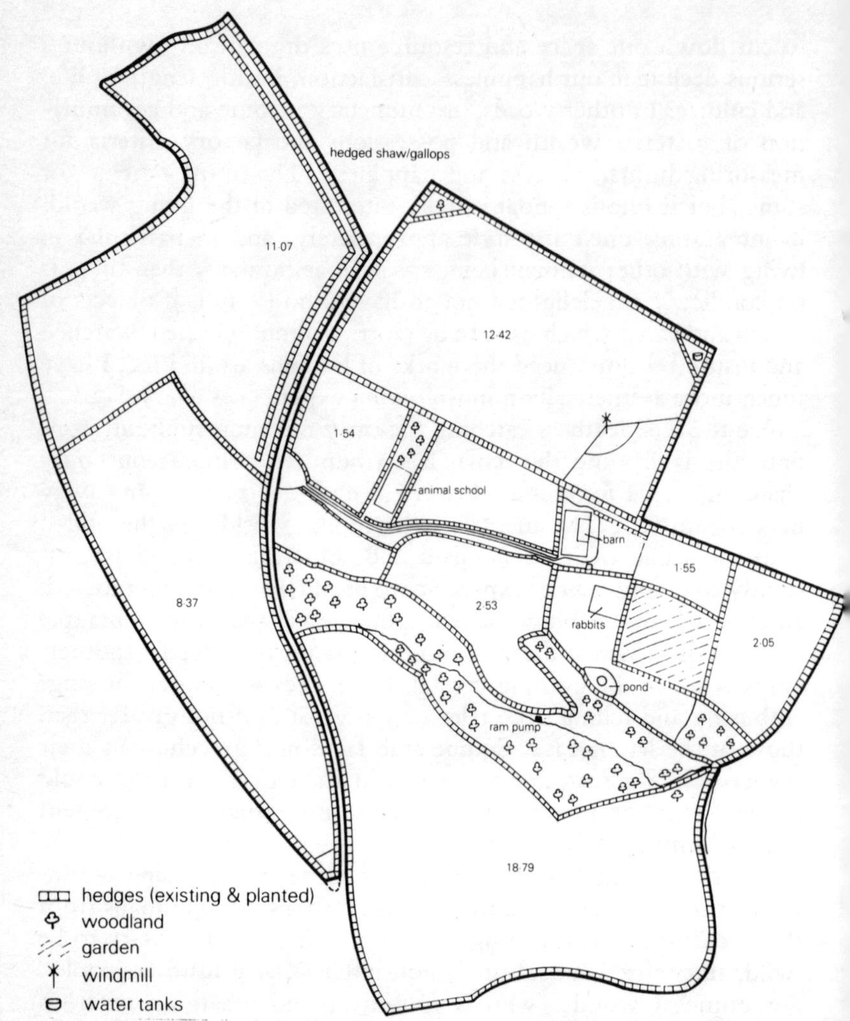

Figure 14. Little Ash Eco-Farm, Throwleigh. Scale 1:25,000.

slaughter. For example, the highland sheep we brought with us from Mull raise their own lambs, but we also milk them once a day and sell sheep cheeses. Their fleeces are kept and spun, and we sell spun wool, kits and designer knitwear from our ecologically raised animal wool.

As usual our South Devons suckle two calves each, thus allowing extra bought-in calves to be raised by mothers rather than in intensive calf-raising systems. We also milk some of our South Devons sometimes, and have some Jerseys for our own cheeses, butter and cream and some yogurt for sale. Not having a milk quota, we are not allowed to sell the cheeses (see pp. 40–1). Why milk quotas are not government property, which must be returned to the ministry when they are no longer required in order to be given out to starter and small-producing farmers, escapes me . . . but then there are so many irrationalities in agriculture it is no longer surprising to find such unfair idiocies operating.

We milk the mares too; of all the domestic animals, their milk most closely resembles human milk. It is useful for premature babies, people who do not want to use powdered milk and do not have enough of their own for their child, and gourmets wanting to make Mongolian horseman dishes.

We also have angora rabbits and llamas, who produce beautiful fleeces which we have spun and which are included in our designer knitwear. Our llamas are delightful companion animals, and we have trained several of them to carry the sandwiches and waterproofs for walkers heading out onto Dartmoor.

Our horses teach people to ride and drive and work horses on the land; they do much of the work of hay-making and maintaining the grassland, and almost all the work in the gardens. They take people on long week treks, provide energy, enjoyment and education for humans, and breed. We sell young educated horses to go on to great things, but only to approved people.

Our animals are, above all, our companions and friends. Our dogs, collect, move and separate other animals, collect and deliver objects and are enthusiastic helpers. Humans and all the others for whom we provide food, lodging and education are expected to contribute in one way or another to the work of maintaining, improving and running the eco-farm.

We have about one acre of garden, including fruit and eventually, when we can afford all the trees, an orchard a mile long. This is along the shaw of a field, and allows for recreation and enjoyment on the farm by allowing people to come and pick fruit, wander around the farm, ride or drive a horse-drawn wagon around the orchard, picnic and give their dog a walk as they go. It is an experiment in multi-purpose land use, integrating recreation, education and food production of the land.

There are many other ideas we are trying out at Little Ash.

Although initially we were regarded with suspicion by the professionals in the National Park and by local agricultural advisers, they are now interested, realising that we may well be finding at least some answers without massive input of government money. The farm already works economically, but it still has a long way to go with developing and putting into practice many of our ideas. I hope we shall have the energy and motivation to continue this work.

8 Animals and Ecological Agriculture: Towards Symbiosis

In chapter 2 we mentioned the major areas for debate regarding keeping and raising farm animals, but we have not yet considered these arguments in detail. I started my first farm because I was concerned about these problems. Over the years and farms we have discussed, tried out, altered and developed alternative appropriate animal husbandry which takes into account the ecological, ethological (animal behavioural) and ethical problems. Three farms and 20 years on, I think we have important contributions to make to this debate, and to help people who are concerned.

Animals kept under almost all husbandry systems, particularly those in intensive animal housing, in laboratories and zoos, and companion and sport animals, are often severely restricted in the activities they can perform (Harrison, 1964), and have a host of behavioural problems (Campbell, 1975; Kiley-Worthington, 1977). In addition, keeping animals of any type for any reason will have environmental effects which can be undesirable—for example, using resources and producing waste products.

The attitudes of human beings to even one group of animals, mammals, are shrouded in ignorance and full of anomalies, inconsistencies and irrationalities. For example, they will be appalled at the intensive conditions in which many pigs and calves are raised and kept; after being shown such a unit they may even, for a while, give up eating meat from animals raised in such environments. The same people will make a hero out of an Olympic gold medal rider, although his horse will probably have been bred and raised in what, to the horse at least, may be considered similar conditions. He will probably have been conceived using severe restraint on the mare, often with manipulation of genitalia and drug treatment. He will more than likely have been isolated from his mother at the age of three to four months, and thereafter raised in restricted conditions, isolated at least partially from his kind, unable to have normal social intercourse with other horses, move freely around, eat at will, or perform many behaviours in his repertoire. Even the normal way he would spend his time has been grossly interfered with by feeding him a diet restricted in fibre but high in nutrients, in order to make him grow bigger and, so the belief goes, perform better. As a result he suffers increased risk of disease and in particular digestive problems. He may have suffered intrusive surgery, ranging from castration to denervation of the lower legs to prevent his feeling lameness; or he may be on analgesic drugs to prevent him feeling pain in his overstrained legs. Finally, as a result of this 'stressful' environment, he may develop behavioural problems such as aggressiveness and neurotic or stereotyped behaviour (Kiley-Worthington, 1977 and 1987b). To prevent the latter he may have had major surgery. All such practices are legal and acceptable by the international horse establishment and Olympic committees.

However, in many Western European countries, such as West Germany and Switzerland, the raising of food animals under similar conditions is either illegal or the subject of current debate.

An alternative, more rational attitude to animals and their keeping is necessary. Few people seriously concerned with animal welfare issues dispute this. However, controversies do arise over where the limits to human manipulation of animals should be drawn. At present there are several extreme positions held: from those who believe that there should be practically no interference with animals whatever (e.g. Reagan, 1983), to those who consider that there should be almost no restrictions on the way they are used or kept (Frey, 1983; Paton, 1984).

The aim of this chapter is not to examine in depth all these various arguments, which has already been done many times (e.g. Clark, 1976; Singer, 1976; Rollin, 1981 and 1989; Midgeley, 1983; Reagan, 1983; Saponsis, 1987), but rather to argue for a blueprint which may be helpful in deciding where the lines should initially be drawn between the use and abuse of mammals and birds.

The intention is to produce a pragmatic, realistic set of guidelines which can be adapted to any animal husbandry system but which concentrates here on farm animals.

ECOLOGICALLY SOUND ENVIRONMENTS FOR ANIMALS

There are two initial points which are relevant here:

1. When designing environments for animals, their natural habitats should be considered and an effort made to try and re-create them because they are the environments the animals have evolved to live in and are therefore best adapted to.
2. We must look at the global as well as the local implications for the development of animal husbandry systems (for example, intensive pig and poultry houses), a subject that has received little attention.

As we have already seen (p. 57), from an ecological point of view all living things are interrelated with each other and with the environment. What separates one ecosystem from another is that the mutually dependent relationships between the species within the ecosystem are stronger than those between ecosystems. In addition, as we have previously stressed, ecosystems tend to be self-sustaining (p. 58).

For the environment of our animals to be ecologically acceptable, therefore, these and other factors should be considered. For example, what are the environmental effects, locally and globally, of the husbandry of a particular animal or group of animals? Where does its food come from, or the materials used for providing shelter, heating, and so on, and what are the environmental effects of these?

In addition, the animals must not endanger the survival of other species in the habitat, for example by transferring disease, or create a nuisance to other species by polluting rivers with their sewage, aggressing others persistently, creating ugly, smelly environments and so on.

One example of the local and global consequences of an animal husbandry system is an intensive broiler farm. These units house tens of thousands of birds that have been selectively bred to grow extremely fast and at best have a conversion rate of plant to protein of 2:1. It is usually considered that they provide cheap, desirable food for many people and are very 'efficient' protein-producing machines.

However, they give rise to an overriding problem with manure disposal, since large quantities of manure applied to the land result in pollution of the water, and an excess of nitrogen in rivers, lakes and canals. They cause smells and flies, the buildings are usually aesthetically unattractive, and many people are upset by the thought of thousands of birds being raised in enclosed, ugly buildings under confined, crowded conditions with very considerable behavioural deprivation. Such animal husbandry systems make undesirable neighbours for humans and are often now hidden in relatively isolated places.

For the chickens to grow fast and survive they are given feedstuffs high in proteins, often imported from other countries (see chapter 2). The depletion of fish stocks (sold for fish meal for these chickens) due to over-fishing can cause other ecological changes, sometimes on a large scale.

In order to grow quickly, the chickens must be housed in environmentally controlled buildings which require considerable resources in terms of the materials used for their construction—sometimes imported hardwood from tropical forests, or softwoods which have been planted over vast areas of Europe in order to cope with consumer demand, and can reduce species diversity and the beauty of the countryside (Shoard, 1980). They also use energy, often generated from non-renewable resources.

The physical health of the broilers is maintained only by extremely strict inoculation procedures, sterilisation of living areas, and drug treatments sometimes including in-feed antibiotics.

There are also considerable economic costs, and if all these costs are added up, the net biological production is not necessarily more efficient than that often despised egg and chicken meat producer, the backyard hen (Blaxter, 1976, and chapter 2).

Ecologically and ethologically sound chicken raising results in smaller numbers of free range hens who occupy the 'niche' on the farm which they have evolved to live in: they live on waste products from other animals, plants and decomposers, and on food produced on the farm. They can be managed to glean after arable crops, living in houses constructed from local materials, and moving around the farm (Balfour, 1975; UFAW, 1981). It is true that they would be fewer in number, and would grow more slowly; however, their net biological cost of production would be less, the environmental effects drastically reduced and, it is maintained by some, they taste better and are less likely to poison the consumer. The question the consumer must ask herself is whether the 'luxury' of eating chicken daily is worth all these environmental consequences, since it is not necessary for her health.

I was recently sent a sales brochure for 'organic fertiliser'. I telephoned the producer to find out what it was made from, and was told it was largely 'free range chicken manure', yet he was clearly producing many thousands of tons of it. Yes, he said proudly, he was the biggest 'free range organic chicken producer'; he had tens of thousands of chickens. I asked how he could have so many free range, and he said it was easy: they were in pens, and various constituents of their ration were of course bought in, not all from Britain. If they are in pens, whatever else they are they are not free range. An ecologically sound environment means that no species will be kept in very large numbers, since the populations will only be able to grow to the point where they can be sustained by the diversified system *ad infinitum*.

The more diverse the system in terms of the number and types of species within it, the less likely it is to succumb to large changes. In the last 15 years, the importance of endangered species and active interest in conserving them has risen dramatically. The usual reason for this interest is aesthetic (it is a shame to lose irreplaceable and beautiful species). This is a serious consideration (e.g. Rolston, 1983), but there is also a vested interest in conserving species in order to maintain the stability of the biosphere. As Erlich and Erlich (1982) put it, loss of a species is like 'rivet popping': it weakens the biospherical structure.

Applying the concept of diversity to agriculture leads away from specialised monocultures towards mixed farming and the conservation of many different habitats on the farm. Pigs and chickens, instead of being housed in large numbers in purpose-built intensive housing systems, and fed inappropriate foods from off the system, are treated more like the scavengers they are in the natural ecosystem, and are kept in small groups.

Thus the animals occupy their 'niche' or a close approximation of it.

Even if the animals are allowed to roam over vast areas, their keeping may not be ecologically acceptable. Shane (1987) eloquently points out the role of beef-eating patterns in the United States in the destruction of South American forests, and the worldwide ecological effects of this. The same has been true of vast regions of the Third World for some two to three decades (e.g. Dumont and Rosier, 1969).

If they cause problems like these, animal husbandry systems cannot be considered ecologically sound environments and are therefore unacceptable, even if the animals in them are well fed, well kept and experiencing 'good welfare'.

These ecological concerns are not usually raised in welfare debates. Perhaps it is time that 'welfare' took into account the environmental and ecological effects both globally and locally, as well as considering both the individual and the species or group, before coming to conclusions about what is right and what is wrong. Up to now, people concerned with animal welfare as a rule are interested in, and give priorities to, individuals, whereas the environmentalists prioritise the species or group. The result has been conflicting moral codes which have led to divisions and confusions in management practices between these two groups. Time is growing short on the environmental front, and it would seem sensible to try and integrate the two and balance up the interests of all relevant parties for each decision.

It is true, I am assuming here, that life itself is important. For example, it has been argued that if there is no life there can be no suffering. The RSPCA took this position concerning foxes in 1976; they did not mind if there were no foxes, but if there were any then they should not suffer. However, the absence of foxes will have effects on the biosphere as a whole because of the interrelatedness of all. What these effects will be we do not exactly know, but as ecologists we do know that, as Erlich and Erlich say, the loss of each species will fractionally weaken the structure of the world.

It is as well to remember that neither an individual nor a species can exist independently of its environment; these ecological considerations are concerned with the animals' or the species' relationship with the environment. They are consequently *fundamental* to any debate on welfare: the quality of life of any being including humans.

Aesthetic Considerations

We must not ignore aesthetic considerations in the design and management of animal husbandry (Shoard, 1980). But the aesthetics of animal husbandry systems are complicated by the possibility that the animals themselves may have some form of aesthetic appreciation or objection. For example, horses appear to dislike strong human smells, and to like and choose to be where they can have a good view. A distinction here between uncomfortable or deprived environments and aesthetically unpleasing ones for the animals is difficult. This is not the place to take this further, but it is a possibility that cannot be ignored.

Table 18 lists the various concerns that must be taken into account before the animals' environment will be considered ecologically sound.

Table 18. Criteria for ecologically sound environments for animals and humans

They should cause no long term or irreversible environmental change by considering the local and global environmental effect of all aspects of the husbandry. In particular:
1) The effect on other species of plants and animals.
2) The long term and short term effects on the physical environment, e.g. soils, tree destruction, etc.
3) The effects on local humans of the husbandry (e.g. any 'nuisance' or environmental value).
4) Provision of appropriate food which causes no adverse ecological effect locally or globally.
5) Provision of other environmental needs of the animal. For example, supply of materials for shelter, shade, nesting materials, heating, etc. and their environmental effect.
6) Appropriate climate and ability to adapt to changes.
7) The origin of the animal, and its local and global effect (particularly if captured from the wild).

ETHOLOGICALLY SOUND ENVIRONMENTS FOR ANIMALS

Ethologically sound environments are those which reduce or eliminate prolonged animal suffering. Many environments that have been designed for animals in the last few decades have concentrated on meeting the requirements of the human handlers, rather than giving first priority to the physical and psychological 'needs' of the animals housed in them.

That animals feel pain, can suffer, and that they feel emotions—

terror, fear, happiness and joy among others, albeit that these states may be somewhat different from those typically experienced by human beings—is now generally agreed by ethologists seriously concerned with animal welfare (Stamp-Dawkins, 1980), as well as by philosophers (Rollin, 1981 and 1989; Midgeley, 1983; Clark, 1976; Singer, 1976) and even psychologists (Walker, 1983; Dickinson, 1980; Pearce, 1987). The problems arise in defining for each species and individual what constitutes prolonged suffering, and what are its behavioural needs.

Suffering

It has been assumed that suffering results from 'stress'. This is widely used as a physiological term and involves in the first instance adaptive responses which allow an animal to withstand environmental changes (Selye, 1952). Short-term stress responses are therefore desirable and not consequently indicative of suffering. However, should this state continue, then these bodily responses may become destructive to health and happiness. Ethologically sound environments are therefore concerned primarily with avoiding prolonged stress. It is also worth considering that experiencing these states is often relative. For example, if it were not possible to experience fear, it would not be possible to experience its absence which can give rise to positive emotional experiences such as 'happiness' or 'joy'.

It is often argued (e.g. Beilharz, 1988) that by breeding selectively for tolerance to particular conditions, and keeping the resulting animals in changeless, comfortable environments, we could breed animals which would not suffer, and of course by implication would not feel emotions. The snag here is that unless the nervous system were completely desensitised, the threshold for responses by these animals would drop, and the animal would then react violently and emotionally even to very slight environmental changes. A better solution might be simply to grow protoplasm and animal proteins in test tubes.

Distress

Behavioural evidence for unpleasant emotional states is called 'distress'. It is a useful term as it avoids the often unknown physiological changes which are implied by the use of 'stress', for which there may be no direct evidence. An animal is described as 'distressed' from observations of his behaviour and his environment. If he is behaving in a distressed way, then we can assume that he is likely to be suffering. Whether or not we have demonstrated a relationship between these behaviours and prolonged physiological stress, we give the animal the benefit of the doubt, which I believe we should.

We can also assess possible *behavioural distress* that is possible prolonged psychological suffering. For example, abnormal behaviour and apparently purposeless or self-destructive behaviours such as stereotypes (repeated actions like head-swaying in institutionalised humans), abnormal aggression and cannibalism can be used as indices of *unhealthy* animals (Kiley-Worthington, 1977; Jensen, 1986). Neuroses and pathologies also tell us that all cannot be well.

But are there other ways of identifying behavioural distress? After all, a prisoner confined to an isolation cell may not initially behave neurotically or pathologically, but he may well be distressed for prolonged periods. To try to answer that question, I studied confined, crated calves raised for veal, and compared their behaviour with that of mother-reared free-range calves (Kiley-Worthington, 1983). Few of the veal calves behaved very obviously peculiarly, but on analysing all the figures, I found there were some very important differences (Table 19) showing that other criteria must be taken into account. Further study with single-stabled horses (Table 20), compared with those in groups outside and with feral horses, and other workers' comparisons with different groups of pigs and dogs, indicate that we can use these criteria to help us measure distress, as evidence that the husbandry system is wrong:

1 *Physical ill health*

This is the criterion usually used by the public and veterinarians to indicate physical suffering and distress. There may be malnutrition, diseases left unattended, problems resulting from lack of appropriate attention from humans, such as foot-cutting in horses and donkeys.

2 *Occupational disease*

This includes disease and physical damage as a result of husbandry and performance—for example frequent lameness in dogs and horses whose job is to run far and fast or jump high; fertility

Table 19. Differences in behaviour between calves raised in crates for veal and mother-reared at grass in 41h (adapted from Kiley-Worthington, 1983 and Kiley-Worthington and de la Plain, 1983)

	Confined calves	*Mother reared calves*
Number calves:	12	8
Total number calf hrs observed and recorded	1,152	20,800
Activity minutes/hr:		
Lie	39.3	32*
Stand	4.7	8*
Eat	3.5†	6.1*
Move	1.3†	2.8*
Ruminate	□ 7.8	5.2*
Self groom	□ 6.2*	3.0
Investigate	0†	3.2*
Chew self or objects	□ 5.1*	0
Suck mother	0†	3.5*
Play and other activities	0†	7.6*
Social contact times/hr	0.7†	1.2

* = significant difference between two populations.
† = these activities controlled or impossible in confined calves.
□ = abnormally high levels: abnormal behaviours, stereotyped or self destructive in some cases.

problems in dairy cattle who do not have access to a bull, and so on (see also 3 below).

3 *The persistent need for surgery and drugs to keep the system operative*

SURGERY. Surgery, other than for life-saving medical reasons, includes:

i) Cosmetic surgery (such as tail-docking and ear-clipping in dogs).
ii) Surgery to compensate for over-use or misuse of limbs or other parts of the animal (denervating of the lower legs of many competitive horses to eliminate signs of lameness).
iii) Surgery to correct faults in the anatomy or physiology of an animal that has been bred to have certain characteristics considered

Table 20. Indications of distress in stabled horses

	Racing stables		*Teaching stables*	
No. of establishments	5		12	
No. of horses	76		150	
	number	%	number	%
No. on Butezanodole	5	6.5	43	28
Denervated	20	26.5	6	4
TOTAL	25	33	49	32
Wood chewing	70	92	50	33
Crib biting or wind sucking	6	7.8	9	6
Weaving or stable walking	12	15.7	10	6.6
Head throwing or tossing	10	13.1	20	13.3
Stable kicking	8	10.7	15	10
High aggression levels	30	39.4	35	23.3
Stable neurosis	26	34	47	31.3
TOTAL	162	212.7	186	123.5

important for the breed—for example, eye surgery for Pekingese dogs, nose surgery for dogs with squashed noses such as bulldogs and boxers, hip surgery for German shepherd dogs whose breed characteristic has been to have slouched, low hips. Belgian Blue cattle, as mentioned in chapter 2, have been bred with such enormous hips that they cannot give birth normally, and the calves have to be delivered by caesarean section. The modern white turkeys have been bred to have such large breasts that they are unable to copulate, and this has to be done by artificial insemination.

iv) Surgery to prevent or reduce the performance of some undesirable behaviour (neck surgery in horses to prevent the animals crib-biting, de-barking in dogs, de-clawing in cats, tail-docking in pigs and debeaking in hens and turkeys to prevent cannibalism). The most widespread and frequent surgical practice in this regard is castration of males and sterilisation of females, usually by hysterectomy. If we are seriously interested in the animals' welfare, then we must seriously question whether castration is always necessary. Denial of sex and having offspring is something humans find very unacceptable, so what evidence have we that this is not true of other animals, too? It is possible they value sex and having infants even more than we humans do.

DRUGS. Unacceptable drug uses include:
i) Persistent use to eliminate signs of occupational disease, such as analgesics (pain-killers such as Butazanodole) for lameness in horses, the result of overstraining.
ii) Drugs used as tranquillisers and sedatives used persistently or routinely for behavioural reasons. These are often used to compensate for inappropriate or no handling and training. The use of immobilisers to allow captive animals to be handled is also unacceptable in the vast majority of cases. If animals are appropriately trained and routinely handled by skilled people, this use of drugs is unnecessary; the need for them indicates faults in the handling, training and keeping conditions.
iii) The use of drugs for manipulation of breeding, or to change the animal's physical or behavioural character, such as steroids (hormones) used to synchronise breeding, or to encourage weight gain.
iv) The frequent use of drugs to control persistent disease outbreaks which are the result of inappropriate management and environmental design for the animals—for example, antibiotics fed routinely in the feed. Antibiotics are also commonly used to make animal management systems economical, by curbing disease outbreaks in intensive hen, calf and pig units.

Such surgery and drug treatments are unacceptable in the long term, and in the short term would be acceptable *only* if they prolonged the life of the individual, allowing the animal to continue to live if he would otherwise be killed as a result of his physical or psychological problems. This may be necessary with the present generation, but if we are seriously interested in animal welfare, it is important to breed and manage future generations so that they do not become dependent on such drug treatments for survival. We now have enough knowledge to be able to do this in almost all cases.

4 *Behavioural changes*

There are also behavioural changes that take place in animals (as in humans) when they are distressed. The most important are:
a) The performance of abnormal behaviour which appears to be of little benefit to the animal. Neuroses and pathologies are included here as well as activities such as pacing, or running at the bars—behaviours often seen in confined, restrained farm animals.

Other behaviours in this category include intense fear as a result of handling, for example, or a sudden environmental change. Some ruminants 'pseudo-ruminate' when they have no food in their stomach that needs to be ruminated.

b) Stereotypes. These are defined as behaviours which are *constant in form, fixed in all details and apparently purposeless.* They include crib-biting and wind-sucking in horses, bar-biting in some sows, some head tosses and throwing around (by cattle, horses, sheep, big cats, elephants and others).

Recently it was found that natural opiates were secreted when pigs were performing a stereotype (Danzer *et al.*, 1983). It was then argued that it was acceptable for the animals to perform such behaviour since it is a way for them of achieving a 'high'. The fact remains, as for institutionalised humans who also develop these behaviours, that if the conditions are so unacceptable that they have to find a way of self-stimulating to escape from them, surely that will not do, from either the animal's or the human's point of view. What it does do is confirm that if stereotypes are performed, then we *know* that the animal is distressed: the environment is *not* ethologically acceptable.

c) Substantial increases in aggression compared to feral or wild members of the same species. We know that aggression tends to increase in animals and humans that are stressed for prolonged periods. This may be because they are crowded, restricted and/or frustrated, as well as for other reasons (Ulrich, 1966).

d) Frustration is also often associated with prolonged distress. If, therefore, we find that there are high levels of behaviours that are usually associated with frustration in the animals we are assessing, then we will know that they are likely to be distressed. A great increase in behaviours such as pacing, head-shaking, pawing, head-tossing and self-grooming are some indicators of this (Duncan, 1978).

e) It is important to see how the animals allot their time. Have their 'time budgets' changed dramatically from those of wild or feral animals, and if so why? Horses, for example, when at pasture spend around 16 hours a day eating; when confined to individual stables and fed restricted amounts of high fibre food, they spend much more time standing doing nothing and less time eating (Kiley-Worthington, 1987b; see also Table 21, pp. 182–4). Pigs and humans are reported to increase the time they spend sleeping when confined and unable to perform many other activities. In such cases, abnormal behaviours may occur to fill up the 'spare time'. Animals in the wild of course also show changes in time budgets dependent on day length, available food, or their success in catching prey. It is only large time-budget changes accompanied by restrictions in behavioural options that may be indicative of distress.

f) Substantial changes in how behaviour develops during the animal's life can also be indicative of prolonged distress. For example, calves which are shut in crates from 2–16 weeks old walk when released as if they were two days old (Kiley-Worthington, 1983).

These are the most important indicators of prolonged distress. However, this is not to say that if *at any time* any of these criteria are fulfilled, the animal is distressed and therefore his environment should not be considered appropriate. A certain amount of stress and distress is inevitable and perhaps even desirable throughout a being's life. On the other hand, if several of these criteria are fulfilled and occur frequently, then there are indicators that the environment is not acceptable to the animal—he is distressed and changes must be made.

There is another very important way of assessing the animal's (or human's) welfare, which is often done intuitively when we talk about an environment that is 'cruel'.

5 *Behavioural restriction and psychological needs*

Is a dog who is tied up 24 hours a day 'cruelly' kept? Should a sow be confined to a stall all the time, or a horse in a stable? What these conditions have in common is that they restrict the individual's behaviour in many ways. So far, 'behavioural restriction' and ways of measuring it in different environments in order to be able to assess their relative suitability have not been seriously considered by other scientists interested in animal welfare.

Thorpe (1965) considered that an animal should be able to perform the normal behaviours in his repertoire, but he did not expand further on this. In making decisions concerning ethologically sound environments for animals, it is essential to have some understanding of their 'behavioural needs' and how to assess them. This debate has become more active in the last decade (e.g. Hughes and Duncan, 1981).

To try to answer these questions, Stamp-Dawkins (1983) pioneered choice tests. Here, the animal is given a choice between different environments and 'asked' to vote with his feet. The most important objection to choice tests is that such an approach ignores the fact that we already have some knowledge about the animal and where he lives, what he does and so on, from the evidence of evolution. All the behaviours he has in his repertoire have a function, they have been selected for by evolution for some specific reason. Thus we already know that preventing him doing them will not be in line with how he has evolved to behave and will therefore be more likely to cause him distress.

It is of course true that the animal (or human) cannot and does not do everything at once, that from moment to moment there are different behavioural priorities. However, because a behaviour happens very rarely, or takes up very little time, does not mean that it is not necessary, or is a luxury. Mating, for example, does not take up a great deal of time or happen very frequently for most mammals, yet it is recognised as being *essential* for the survival of the species, just as eating is. In humans it is also recognised as being a great hardship to be unable ever to mate.

Inevitably, then, if we take evolutionary ideas seriously, we already have a blue-print on how to optimise an animal's environment so that prolonged suffering is reduced or eliminated. This will be one in which he is able to perform *all* the behaviours in his repertoire. We do not need to carry out choice tests, but we may need more information on his natural, unrestricted behaviour.

However, certain behaviours, for example predators hunting and killing, will cause prolonged or severe suffering to others who have an equal right, while under the jurisdiction of humans, to a life of some quality. Thus we may perhaps suggest that in order to reduce or eliminate prolonged or acute suffering to those animals, it might be possible '*for the animal to perform all the behaviour in his repertoire, provided this does not cause prolonged or acute suffering*'.

Table 21 assesses the behavioural restrictions for two farm animals, and dogs.

In the last two decades we have accumulated more and more detailed information on what the different species do and how they organise their societies, and we do have sufficient information to be able to design their environments to reduce distress.

Thus, if we are to be concerned not only that an animal should stay alive but that he should have some quality of life, then we

Table 21. The behavioural restrictions for a) hens, b) dairy cows and c) dogs kept in different ways, compared to feral animals

a) Hens

0 = none, or very rare restriction
+ = occasional restriction
++ = usual restriction for majority of the day
+++ = total restriction at all times

	1 *Feral*	2 *Pen + run*	3 *Aviary*	4 *Deep litter*	5 *Battery*	6 *Free range + cock*
Movement anywhere	0	+	+	++	++	0
All gaits	0	0	0	+	++	0
Preening/wing flap	0	0	0	+	++	0
Free social contact	0	+	+	+	++	0
Sexual behaviour	0	0	++	++	++	0
Maternal	0	++	++	++	++	0
Scratching	0	+	++	++	++	0
Environmental change	0	+	++	++	+++	0
Total	0	6	10	13	17	0
Food and water always available	+	+	+	+	+	0
Shelter	0	0	0	+	+	0
Overall total	1	7	11	15	19	0

6 our own hens.

Table 21. (*Contd.*)

b) Dairy cows
0 = none, or very rare restriction
+ = occasional restriction
++ = usual restriction for majority of the day
+++ = total restriction at all times

	1 Feral/ wild	*2 Pasture and shelter*	*3 Straw yard and run out + calves*	*4 Straw yards no run – calves*	*5 Cubicle*	*6 Yoked*
Movement unrestricted	0	+	+	+	++	+++
Never unenclosed	0	+	+	+	++	++
Comfort	0	0	0	+	+++	+++
Easy self-groom and maintain	0	0	0	0	+	++
Choose social partner	0	0	0	0	+	+++
Mixed sex and age	0	+	+	++	++	++
Sexual behaviour	0	++	0	++	++	++
Maternal behaviour	0	++	0	++	++	++
Dull environment	0	+	+	++	++	+++
All gaits and exercise	0	0	0	+	++	+++
Total:	0	8	4	12	19	25
Always find shelter	+	0	0	0	0	0
water,	+	0	0	0	0	0
food	+	+	+	+	+	+
Total:	3	9	5	13	20	26

1,2,3 mixed age groups with at least 1 bull.
3 Our own cattle.

Table 21. (*Contd.*)

c) Dogs

0 = none, or very rare restriction
+ = occasional restriction
++ = usual restriction for majority of the day

	1 Feral	*2 Urban feral*	*3 Rural pet*	*4 Urban pet*	*5 Working*	*6 Breeding*	*7 Show*	*8 Experim.*	*9 Circus*
Movement	0	0	0	0	0	+	++	++	++
Gait	0	0	+	+	+	++	++	++	++
Grooming unrestricted	0	0	0	0	0	0	+	+	0
Social contact	0	+	+	++	++	++	++	++	+
Sexual behaviour	0	0	+	++	++	++	++	++	+
Maternal behaviour	0	0	0	0	0	0	0	+	0
Social hunting	0	0	+	++	+	++	++	++	++
Environmental stimulation	0	0	0	+	0	++	++	++	+
Total	0	1	4	8	6	11	13	14	9
Food and water always available	++	+	0	0	0	0	0	0	0
Shelter from temperature extremes	++	++	0	0	0	0	0	0	0
Possible important social relationships with humans	++	++	0	0	0	0	0	+	0
Possible intellectual stimuli through training	+	+	0	0	0	+	+	+	0
OVERALL TOTAL Restriction quotient	7	7	4	8	6	12	14	16	9

CONDITIONS
Breeding = Dogs in breeding kennels
Show = Dogs kept in kennels predominantly for show
Experim. = Experimental dogs kept in laboratory conditions.

already have sufficient information to design his environment. In the future we may learn more, but right now we should at least use our existing knowledge.

This does not mean that we have to open all the gates and have nothing more to do with animals on our ecological farm, or anywhere else for that matter; what it does mean is that we already have enough knowledge to be able to improve the environments for most of our animals, and we do know how to assess whether or not we have done this. Invention and substitution can often be used to fulfil these criteria, in just the same way as some wild animals and humans use them in adapting to new environments.

There is, however, a body of opinion (e.g. Beilharz, 1988) which considers that domestic animals have changed so grossly during domestication that we can no longer assume that they have the same behavioural needs as their wild ancestors. What evidence is there for this?

The Behavioural Effects of Domestication

There is in fact very little evidence that the behavioural repertoire, and particularly social behaviour such as social organisation, mother and young relationships, communications, sexual and maternal behaviour, has changed genetically during the process of domestication. Given the opportunity, all our domestic animals still do the same things as their wild ancestors. Pigs, although confined to sties for generations with no possibility of rooting, will still root when given the opportunity; domestic dogs still hunt socially when feral, although they may never have had the opportunity of doing this during the 8,000 or so years of domestication, and until they learn how, may be bad at it! Studies on domestic dogs that have gone feral in the United States, Italy, and other parts of Europe, on feral pigs in the United States and Europe, and on cattle and horses confirm this (Stolba, 1982; Kiley-Worthington, 1983 and 1987b. See also Table 22).

An evolutionary biologist does not find this surprising as, in general, there has not been sufficient time during domestication for the necessary genetic changes to have occurred unless there has been very strong selection for particular behavioural changes. With few exceptions (one of these is a successful selection against broodiness in some strains of hen), behavioural selection by humans has been by *default*. By contrast, morphological (body) change has had

Table 22. The genetic and environmental effects of domestication on behaviour of some farm animal species.

Evidence for *genetic* change in behaviour from studies comparing wild members of the same or closely related species (if no longer extant) and feral animals (animals that have been allowed to go wild after domestication). (See Kiley-Worthington, 1977, 1983, 1987b).1990, for refs.)

	Horse	*Cattle*	*Sheep*	*Goat*	*Pig*	*Dog*	*Cat*
Habitat preferences	No change ———						
Home area size	No change ———						
Territoriality	No change if there ———						
Movement around home areas	No change ———						
Social grouping	No change ———						
Maternal behaviour	No change ———						
and association with young	No change ———						
Courtship and sexual behaviour	No change ———						
Aggression	No change ———						
Affiliation	No change ———						
Communication	No change ———						
Food preferences	No change ———						

Evidence for the effects of *individual lifetime experience* causing change in behaviours

	Horse	*Cattle*	*Sheep*	*Goat*	*Pig*	*Dog*	*Cat*
All above can be changed in one way or another for all species ———							

In addition:

A) Association with human and attitude to human can be changed by early experience (imprinting) and by experiences later in life (e.g. adult animals tamed and then associating closely with humans, or adult animals previously associating with humans having bad experiences and then avoiding them, or becoming aggressive).

B) There is no concrete evidence that if any of these species are born feral and then caught, trained and tamed by humans, they will be more adaptable, and easier to manipulate than traditionally wild animals born and raised in captivity. This evidence is crucial before we can assume domestication has changed behaviour dramatically *genetically*.

The animals' behaviours, including human animals, are the result of both their genetic constitution and their lifetime experiences.
1) Fundamental behaviour has changed GENETICALLY very little if at all as a result of domestication.
2) LIFETIME EXPERIENCES cause dramatic changes even in fundamental behaviours, in all mammals, including humans.

a conscious artificial selection towards a defined goal (such as larger back ends in cattle for more expensive cuts of meat; or larger litter size and quicker growth of pigs, squashed noses in Boxer dogs, or coat length and colour in cats).

On the other hand, throughout the 8,000 or so years of domestication people have, usually unconsciously, selected individual animals that behaved appropriately towards them. The types of behaviours that have been generally selected for are:

1 Lack of timidity towards humans. The degree of genetic change here is probably small. The effect of the individual's past experience seems to be crucial since wild animals can be easily tamed, particularly if taken from the wild when young.
2 Individuals that are less aggressive to humans may have been selected. However, the present debate on breeds of dog, in which it is maintained certain breeds are genetically aggressive, needs proper research. The answer is that we really don't know if this is genetic or the result of their life-time experiences (that is, the way they have been taught to behave).
3 Animals who were willing to *co-operate* with humans.
4 It is possible that the speed of learning and ability to learn human communication systems has also been selected for.

One of the problems with assessing differences in behaviour, learning, and so on between domestic animals and their wild close relatives or conspecifics is the preconceived ideas of the handlers and thus the way the animals are treated. There is little scientifically tested information on this. In one of the few studies examining possible differences in training and problem-solving between wolf and malamute pups (Frank and Frank, 1982; Frank *et al.*, 1987) it was found that although there were some differences in the wolves' and dogs' pups in the given tests, there were much greater differences in behaviour between the wolf pups socialised to humans and those not. Another important point is that the animal may well know what is required, but is not motivated to perform.

These factors, and others, must of course be taken into account in the design and implementation of management, but they do *not* indicate fundamental genetic behavioural changes or changes in behavioural 'needs' as a result of domestication. We do not have the evidence.

If there is no evidence for fundamental genetically programmed behavioural changes in our domestic animals, how then should we design their environment in order to avoid causing them distress and suffering? The sensible course is to find out how that species has evolved to live, and then to try and allow him to do this as much as we can. This will involve:

1 *Taking account of social organisation*

The animals should be kept in the social groupings in which they associate in the wild, that they have evolved to live in. Different species have different requirements. Thus pigs live in family groups and cats are semi-solitary. Horses and cattle are social herbivores but have rather different social organisations—for example, horses live in family groups which can join together into larger herds, and usually have only one adult male. The strongest bonds appear to be between generations (mother to daughter). In cattle, the bonds between peers—within generations—are very strong, and the family/friend groups live with several bulls. This means that we should have different management for cattle and horses: horses should not be kept in peer groups and only one stallion with the mares; cattle can have several bulls with cows.

2 *Taking account of habitat*

The place where the animals are kept should approximate the type of habitat they have evolved to live in. For example, if forest animals, living in small family groups, are forced to live in large groups in open environments with little cover, it is likely to cause them trauma. With thought it is almost always possible to cater for the species' demands in this way.

3 *Taking account of the natural food, its nutrient content and textures*
The food animals are fed must also be appropriate. Cellulose converters have evolved to convert cellulose and they should be fed diets high in cellulose. If they are not, physical and psychological problems can result—for example, colic and azoturia in horses, reduced life expectancy in cattle (Kiley-Worthington, 1986 and 1987b).

4 *Allowing for the performance of the entire behavioural repertoire which enhances survival, and does not cause suffering to others*
How can this be done? Examples are given in the next chapter.

There are other species characteristics which must also be taken into account. These are:

5 *The way the animals perceive the world*
This can be assessed by studying:
a) The species-specific characteristics of the sensory receptors (the eyes, and the animals' detection of smell, touch, taste, and hearing). For example, the horse's eye differs from the human's in its position on the head which controls the field of vision (Figure 15); the pig's snout is very sensitive to smells. What 'view of the world', then, do the horse and the pig have, which differs from the human's?
b) The species-specific brain anatomy. This can tell us something about the relative importance of different brain functions for the species. Here we need much more information on species-specific differences. Nevertheless, there are differences in the size of different parts of the brain, and in the overall size of the brain, between species, although exactly what this means is not at all clear (Jerison, 1973). The mutilation or sacrifice of many animals in order to find out more about such questions would not be ethically desirable, or ethologically sound. There is great scope for further anatomical studies on animals already slaughtered, and much more could be discovered using simple behavioural techniques such as discrimination tests. (See Figure 16.)
c) The species communication system must be understood in detail. This is particularly important for animals that are trained and are used to help humans with tasks, or perform in one way or another, such as guide dogs, sheep dogs, working horses and dairy cows. For successful working relationships between species

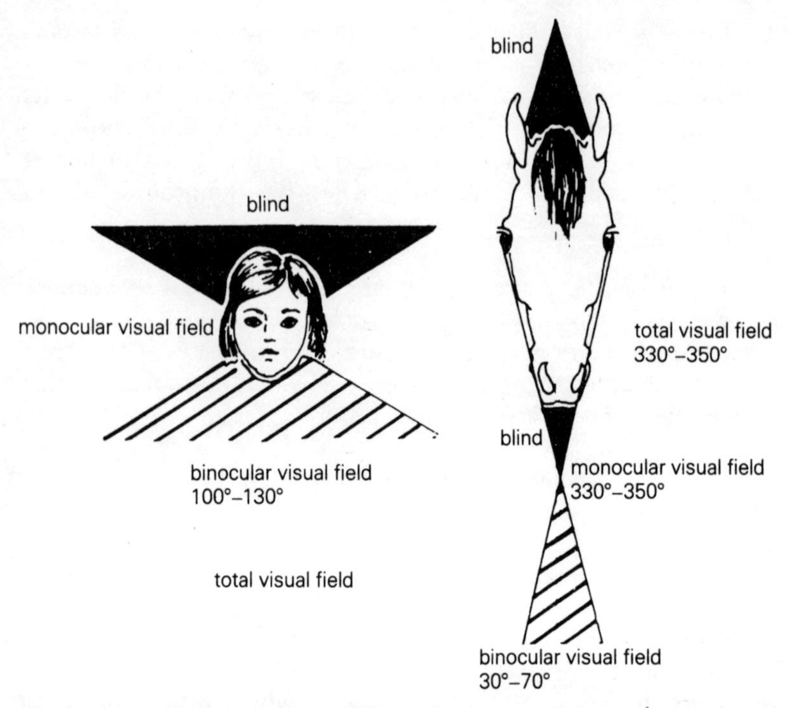

Figure 15. The visual fields of horse and human contrasted. Note the much greater binocular field of vision for a human, but the much more restricted field of monocular vision. Some horses can see right round to the person on their back if they put their head up. (From Kiley-Worthington, 1987b.)

(humans and others) reciprocal animal/human communication is central. Proper knowledge of inter- and intra-specific communication is also essential for the good husbandry of any animal. There can be taught, although some humans will be better than others at this, as in everything else.

Individual animals have their own personalities, but until the species communication is reasonably well known it is often very difficult to understand this. For stockpeople to relate well to animals, they need to be aware of this but they must not have very strongly held opinions of each animal; rather they should keep an open mind to learn.

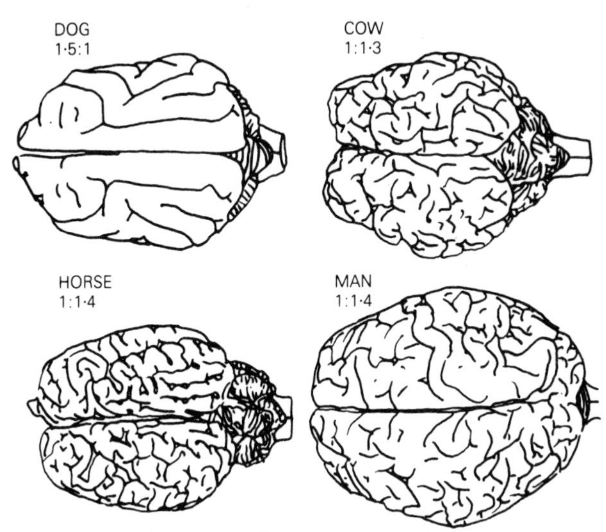

Figure 16. Dorsal view of the brains of dog, cow, horse and human. Ratios give scale.

6 *A good grasp of how animals learn, what each species might be expected to learn, what they have learnt and know, and how they think*

This is particularly important if there are to be good relations between humans and animals, which is the idea of animal management on an ecological farm. Learning plays an important part, even in the everyday husbandry of large numbers of food animals. Appropriate learnt responses can enormously simplify husbandry, and reduce trauma for the animal. For example, sheep or cattle can be trained to come to an auditory signal, instead of having to be herded or even chased to the desired point. However, what animals learn, and how quickly, varies from species to species. Little is known about this from the scientific literature at present, but there is a considerable amount of knowledge among those looking after and training different species, although they are sometimes rather opinionated (Hearne, 1986).

Most species accumulate considerable knowledge about their habitat, other species, weather conditions and their effects, locations of many different things in the habitat (such as shelter, mates, water, morning sun and so on), food differences, ways of overcoming various problems in the environment, to name only a few.

Like humans, they are learning all sorts of things from the moment they are born. Thus it should come as no surprise that, for example, hill sheep have a great deal of knowledge about their hills, and others introduced there as adults will have trouble surviving without this knowledge, although they will quickly accumulate it. Dairy cattle know all about their husbandry system and milking, and their response varies with their stockman's. Any squabbles he has with his wife will be reflected in their milk yields and performance. Horses quickly learn to identify the novice from the skilled rider even from the ground, and remember veterinarians all their lives!

Different species have particular intellectual abilities; humans, for example, have specialised in manipulation and the development of verbal language, horses are very sophisticated visual communicators, bats use eco-location, electric fish pick up and use electricity. Each species is different and must have different views of the world. There is no reason to believe that one is necessarily 'superior' to another; each species is certainly best at being himself, rather than a horse being a human or a human a bat. There are as many ways of perceiving the world and possibly thinking about it as there are species to perceive it. There is real excitement in trying to learn more about how other animals do perceive the world and how they think, in order to have a small window on the differing interpretations to enrich our own human life.

If we are concerned about living symbiotically with animals, as we are in ecological agriculture, then we must take into account the intellectual cognitive differences between species: the horseness of the horse, or the cowness of the cow, in so far as we have any idea of this . . . which is not much to date.

These, then, are the factors important to *the species* that must be taken into account when designing appropriate ethologically sound environments, but there are also factors concerned with *the individual* that must be considered:

7 *The individual's past experience*

Animals, like humans, become accustomed to particular types of environment. For example, dogs who have always been in urban environments do not immediately respond positively to large open spaces, they may in fact become very frightened. Battery hens have usually never had to compete with many other hens to find food, to obtain shelter and so on; nor have they been able to move around. When they are released from their batteries, they often find the change extremely traumatic and some even collapse from the stress. This is not to say that battery hens' environments are acceptable, nor that they should not be allowed the opportunity to experience the great outdoors, but rather to emphasise the effect of past experience. Good or bad experience in their past will affect the behaviour of the animal or human animal—for example, in dogs, weaning too early gives rise to difficulties in socialising; a horse that has been frightened when loading into a trailer may show great resistance to going in again—quite sensibly!

There is a curious belief that animals who are doing the wrong thing are trying to outdo the human; that inevitably there will be a battle for 'dominance'. This is a particularly silly and destructive idea, the 'don't let him get the better of you' syndrome which often leads to major behavioural problems for both animal and human. The animals and human animals concerned will have learnt to respond in the wrong way. The good teacher sits down and thinks about what has gone wrong and how it can be put right; she does not assume that the pupil is trying to outdo her in some way, and even if she believes this is possible, if she wishes to be a good teacher she must ask herself why, rather than set about her student physically because he will not do his mathematics!

8 *Individual personalities*

Even without an age or sex class, animals, like humans, have individual personalities. This is the result of their genetic inheritance and past experience. There are ways in which we can measure personality differences, just as there are for humans (Kiley-Worthington and de la Plain, 1983; Kiley-Worthington, 1987b).

9 *Pleasure*

If we are able to assess fear, pain and distress in animals, then there must surely be some indicators of pleasure. I think there are. Most of us who have to do with dogs would suggest that it involves leaping around, greeting, rushing, playing, sometimes barking, tail-wagging and so on. Table 23 summarises the behaviours in several species which often, although not always, appear to be related to pleasure. We must of course make sure not only that our animals are not distressed, but also that they show some pleasure and joy. Much remains for further research and individual thought.

All these factors must be considered when designing animal environments and educating animals. If they are, we can consider that the animal's environment is ethologically acceptable and he will not show evidence of prolonged distress: our lives together will be mutually beneficial. The main criteria are summarised in Table 24.

ETHICALLY SOUND ENVIRONMENTS FOR ANIMALS

Opinions about how humans should regard animals and the environment have varied with history. In the early days, during the era of the hunter-gatherers, it appears that people lived very much in a give-and-take situation with Nature, as similar societies do today (Turnbull, 1984). The humans must have a certain respect for the natural order of things and how Nature works, but at the same time they are dominated by it and are aware of this.

The first agriculturalists, it seems, had much the same attitude: a very definite respect, but in addition the beginning of efforts to dominate and twist Nature for their own ends.

Table 23. Assessing pleasure in animals

Good physical health, not too fat or thin
Glossy coat
Alert to change but relaxed species–typical behaviour
Elevated paces and head and tail when interested and excited
Leap and chase around
Play games, species–specific
Touch each other if species–typical
Smell and lick each other if species–typical
Species–typical noises of contact (e.g. elephant rumbling, cat purring, low grunts, nickers, etc.)
Stretch
Relaxed dozing and sleeping when appropriate
Rolling if species–typical (dogs, cats, horses, llamas)
Other species–typical behaviours (e.g. rooting pigs, climbing cats, sniffing and trail following dogs, group galloping in horses, gambolling and cudding in cattle, etc.)

Self-sustaining peasants have retained much of this attitude today, for obvious survival reasons. Nature and animals are respected, sometimes admired or even worshipped. It is understood that in order to survive the human must work with Nature and have some understanding of her controls. *There is no question of complete domination*, but rather a symbiotic relationship.

Among many other populations of humans, particularly urban ones, a war has been declared on Nature since the sixteenth century. This has its origins in part in the Judaeo-Christian and Islamic philosophy of the superiority of humans (Thomas, 1984). The result has been that humans have made real efforts to overcome, subdue and dominate Nature and animals, losing any real respect for other living creatures apart from themselves. With the growth of industrialisation and urbanisation, even of the farm, zoo or wildlife park, the predominant cultural attitude in the West has been, and still is, extremely anthropocentric. Nowhere is this more obvious today than in the conventional teaching of agriculture, veterinary and medical sciences and animal management. This is illustrated, for example, by the French word meaning to farm and cultivate: *exploiter*.

If we really did have dominion over Nature, then we would not suffer from earthquakes, storms and floods, plagues of locusts, diseases and at last even death. Technological man does have a

Table 24. Ethologically sound environments for animals

The criteria that must be taken into account:

1 The animal should be allowed to perform all the behaviour in his repertoire which does not cause prolonged or acute suffering to others
2 The animal should be able to associate in the groups, size and structure appropriate to his species and past experience
3 The animal should be in an appropriate physical environment (e.g. forest or simulated forest if forest dwelling, etc.)
4 There must be no evidence of prolonged distress
5 The animal's 'telos' must be catered for by considering the way he perceives the world, his receptors; his brain anatomy, his cognitive ability, his specific learning abilities and his communication system
6 The animal must be considered not only as a representative of a species, but also as an individual, and his past experience must be assessed in order to design the most appropriate environment for him as a) a member of a species and b) an individual

great environmental effect, but we certainly don't have control and we rarely have much comprehension, particularly of the long-term effects of some of our manipulations, as is becoming abundantly clear. The world is too full of examples, from the Ganges floods to the generation of nuclear waste, from Ethiopia to Brazil.

On the other hand, the radical Respect for Life philosophy (Reagan, 1983), although thought-provoking, is impractical without very massive changes in all aspects of human life-style. It is not possible to survive without destroying life or potential life, every time one eats a cabbage, or even a fruit or nut. A modified edition of this theory has some relevance, however. Routley (1975) suggests that a respect for life stance does not necessarily mean that Nature is sacred, or that all living things have an unqualified right to life.

Others (e.g. Clark, 1976; Midgeley, 1983 and Rollin, 1981) argue convincingly that animals must be objects of moral concern. However, Rollin recognises that the interests of one species do trump others, so that the right to life is not absolute as Reagan maintains. More importantly, though, he points out that the killing of animals or causing them pain and suffering is a *moral* decision.

Surely each individual should have the ability to fulfil his 'telos' (be himself and exercise all the behaviour in his repertoire). So he may serve and be served by others, so long as the relationship remains mutually beneficial . . . symbiotic. Each individual, as well as co-operating and benefiting from the presence of others, also to a degree manipulates and profits from them.

Recently an attitude towards animals and Nature of 'stewardship' has been proposed (Passmore, 1974; Attfield, 1983). An unattractive aspect of this approach is its implication that 'big brother knows best': its notion of control and management and the Judaeo-Christian assumption that human interests will always trump those of other species. The Symbiosis Approach is more acceptable, in which humans again recognise their place in the living world and try to work with it for the mutual advantage of all species.

In the last decade or so, several philosophers who have considered animal rights have argued for what we have already referred to as *N. C. Apartheid*: animals must be left alone in special reserves to do their own thing unmolested, with no possibility of developing relationships with humans or with other animals not in their reserve. Such an approach might suit those whose sole interests are in other humans and their doings, but to those of us who are interested in enjoying the company of animals in whatever capacity, it is as unacceptable as the creation of Bantustans in South Africa to the non-racist.

The close association of some humans with animals, which comes from farming, pet-keeping, sport animals, zoos, and in particular the training of animals for various forms of work (guide dogs for the blind, sheep dogs, stock horses, draught buffalo, timber elephants) including entertainments (such as in circuses), *provided* the animals' physical and behavioural 'needs' are catered for, can enormously increase the quality of life for both humans and, as far as we can see, the animals. This also educates the humans to understand that there are other skilled, able and interesting beings who inhabit the world, as well as themselves, to whom they can relate closely. Why should a human's close relationship with any other animal but a human be a sign of that human's sickness or shortcomings, particularly if this inter-species association is preferred to that with another human?

From the animal's point of view such a relationship may be equally rewarding and exciting. In humans, an increase in the quality of life as a result of education is taken for granted by all human societies, although they may educate their children in very different ways. Why should this not be the case for animals, provided it is done in such a way as to fulfil the above criteria? At least *until proved otherwise*, it would seem rational to consider this to be the case.

Thus there is no rational reason why animals should not be used by humans, and humans used by animals, for many activities *provided* the animals are kept and trained in ethologically and ecologically sound environments. The fact that this is not always the case simply indicates that changes should be made, not that we should ban the use of animals for certain things, or assume that they would be 'better off dead'. We do not generally have this attitude to other humans!

'Good' animal management today is still modelled on the patronising stewardship attitude. Thus it involves much interference 'for the animal's good'. It rests fundamentally on dominion and superiority over animals and Nature, rather than necessary co-existence to the mutual advantage of each. The well-meaning but misplaced parallel in treatment of infants or imbeciles and adult animals is in part responsible for this. Adult animals are *not* children or stupid adults. This is not to say that they should not have equal consideration (Singer, 1976).

In practical terms, then, how are we to conduct our relationships with other animals, and how should we design their environments? Is there a practical alternative that fulfils our criteria and reduces the problems to the individual human, animal and the biosphere as a whole and which will in the long run benefit us individually?

The symbiotic approach proposed here implies a healthy respect for evolution and the complex and intricate workings of the biosphere. It recognises that at present humans are the most manipulative species, but also that they have as yet relatively little comprehension of the workings of the biosphere, or the individual living things. It is a rationalistic holistic approach, but also cautionary. It is wiser not to rock the biospherical or ethological boat. Causing change is more likely to be disadvantageous than advantageous, even though it might be ingenious or amusing and apparently show short-term advantages. It is in the long run more likely than not to be disadvantageous to my own or your survival, and that of my or your offspring.

We are now in a position to design practical housing for most

Table 25. Ethically acceptable environments for animals and humans

1 The local and global ecological effect of the system is considered in relation to the biological, environmental and aesthetic value to humans and other animals.
2 The animal is in the type of environment which is ethologically sound, where he is 'happy' and not showing distress, and able to perform all the behaviour within his repertoire provided this does not cause suffering to others.
3 Consideration to him as a sentient being of moral concern is shown.
4 The animal, human and rest of the environment have a symbiotic relationship, which is of mutual benefit rather than competitive. The relationship of the human to the animal could be considered rather as one of an employee than a tool or slave.

animals with these criteria in mind. There is a growing literature on how to do this for different species (e.g. Stolba, 1982; UFAW, 1981; Kiley-Worthington, 1983, 1986 and 1987b and chapter 9).

The 'right' or ethically acceptable husbandry system for animals must take into account all the factors listed in Table 25.

9 Can Symbiotic Living with Animals be Achieved, and if so, How?

In the last chapter I pointed out the guidelines of our animal management that are rationally defensible if we are concerned about both our environment and the animals' welfare. These guidelines will have seemed very radical to some people and some animal husbandry systems. If they cannot almost be fulfilled in a particular animal husbandry system, then it is justifiable to consider that system unacceptable. It must cease. However, having said that, it is not very easy to fulfil all the criteria, even if one wants to. We have been trying for some twenty years on our various farms and finding out what the difficult and not so difficult aspects are. To a large extent we are now succeeding, but there are still shortfalls. This chapter describes what we have achieved so far.

One of the major problems with this approach with farm animals is that even if one does fulfil all the criteria, at the end of the day the animals are being raised to be killed for human food. There is no way round this for some animals, but the longer I do it the less I like sending the animals off to slaughter. Like most other people, I find that, provided I distance myself from them, do not know the individuals going to slaughter well, then it is not too bad. Nevertheless, is it acceptable to slaughter animals at all, even if the conditions of transport and at the slaughterhouse will cause them no distress and pain? The major problem with maintaining that it is *not* right to slaughter animals at all for food is the consequence.

What would happen if animals were never slaughtered?

1 In a short time there would be too many to be sustained. As a result they would gradually be got rid of.
2 Would there be any animals around at all? Would we not, which I fear, just descend into a totally anthropocentric world, with a few precious nature reserves where we could go and

'view' nature like going to the Tate Galley, and where one would have to queue for tickets?

3 How would their roles in the farm ecosystem be fulfilled if we did not keep them? They use areas we are not able to use for food, and their manure allows the whole system to upgrade if properly managed.
4 We stand to lose if there are no animals around, since we would not be able to learn from them or have exposure to different sentient beings.

Even if we had animals for milk, wool, energy or any other product they can produce that we can have without pain, death or suffering on their part, we would still have to kill some of them. For example, if we want milk we must have young born every year. What happens to all the spare males? If we never kill old or decrepit animals we shall end up with very big populations to be supported on every farm. A similar problem arises with human populations in many parts of the world. There is a case, perhaps, for raising some animals for meat, *provided* their lives are happy and joyful. Let us take the major agricultural animals and discuss the degree it is possible to fulfil the guidelines of the previous chapter.

The first priority of all our animals (including humans) living on the farm is that in one way or another they must 'pay' for themselves. This can be assessed in terms of cash, but more usually it is in terms of the products they produce set against the resources and food that they consume. Large herbivores in particular consume huge amounts of food (about 28 lb or 10 kg per day). This means that there is a very real limit on the number that can be supported. Humans, on the other hand, do not consume as much although they demand a very varied diet. They do, however, make large resource demands of the farm in many other ways (for example, heating, washing, equipment and so on), so likewise, there is a very real limit on the number of humans we can have. It would not take long before too many cows, horses or humans would snarl up the system and the farm would stop working! A balance of different species is what is needed.

Suckler Cattle

In Sussex, when we started keeping cattle we decided that we wanted a dual-purpose breed that we could milk if required but which would also be a good beef breed. They must perform well on a diet low in concentrate foods, they must be long-lived and fertile . . . and they must be beautiful, with delightful personalities. Ideally the breed we selected should be local as they would have adapted to the local conditions. The local Sussex cattle, however, had become very specialised single-suckling beef animals. As is usually the case with minority and rare breeds, they were largely kept for showing and to keep the breed around rather than for their good performance and commercial value. They certainly would not have been worth milking very seriously. The only dual-purpose cattle left in England that were available to us to select from could be narrowed down to the Welsh Blacks, the Shorthorns and the South Devons. We finally decided on the big brown curly-coated South Devons who have outstanding growth rates, high quality beef, and in the 1970s some people were still milking them. They were from an area and a climate not too different from Sussex. We bought two heifers and a bull. Twenty years and two farms later, most of our cows are in some way related to those three. When we took them to Mull our South Devons outbred and outperformed the local suckler breeds on the hill and had no trouble at all surviving after the first year. Now for the first time they are in Devon, and doing very well.

It is relatively easy to fulfil the majority of the ecological and ethological criteria when keeping suckler cattle outdoors. However, the fattening feedlots, where the animals are kept indoors or outside in very crowded conditions, fed low-fibre foods, have no bedding, and very restricted behavioural possibilities, are becoming more numerous in Europe and the United States. It is this type of husbandry that is highly inappropriate for the animals, environmentally and ethologically, and disasters are not uncommon. Recently, when I was in Australia, I heard of a case where 2,000 animals in a feedlot had died due to heat stroke because of insufficient shade. Hormone treatments to make the animals grow faster are routine in countries where they are still allowed.

The point is that it is not an either-or case. We could all still eat some beef without causing all the animals and the environment to suffer so, although perhaps we would not be able to have huge steaks every day—but then our health would improve!

Our animals, and the suckler cattle in many parts of the world, still live outside and have a shelter for the winter. The cows are outside all the time in fields. They suckle their own calves, associate in groups as they wish, run with a bull, sometimes several, since we often have some bull calves from the previous year's calvings. The store cattle (the calves from the year before) are also kept and finished ready for slaughter by us in the same way. These are the animals that are usually placed in feedlots when they are brought in from the pampas, sierras, prairies, the outback or the moors.

All the cattle can easily be fed *ad libitum* on food produced entirely on the farm—usually hay silage, straw and occasionally barley. They are usually enclosed in fields, although in the upland areas and on our farm in the Highlands, the suckler cattle were free to roam wherever they wished. The problem there is they have a long cold winter and on many farms may be short of shelter and food for part of the year. Whether this is better or works for the cattle is a complex question; it cannot be decided necessarily in human terms.

The main problems we have found in fulfilling our own criteria are 1) dehorning, and 2) castrating.

Dehorning

When the cattle are fed from hay racks or silage feeders, if they have horns there is competition and some individuals may not obtain their share of food; they may not be able to enter the shelter and there will be much more roughness, with some injuries in the housed cattle, even if they are not shut in. On Mull we could manage not to dehorn the animals we were going to keep, as they had many acres to wander in and we could separate the food appropriately outside. In Sussex and Devon where they have a barn and are confined in smaller fields, having horns leads to these problems. We do therefore dehorn as on balance we consider that they will benefit more, have equal access to resources and that there will be fewer injuries and less competition. The horn buds are taken out with a local anaesthetic as soon as they can be felt in the calves.

Castrating

We have had little trouble behaviourally with keeping the young bulls uncastrated with the cows and adult bull; however, the heifers have to be kept separate, otherwise they will become pregnant before they are fully grown and they may be going to be sold anyway. We have had considerable problems marketing bull beef, though, through the organic markets. The butchers or chainstores that have 'organic' meat have not yet tumbled to the idea that perhaps the animal should not only not be fed chemicals that might harm the consumer, but should live a fulfilling life with minimum behavioural and surgical intervention. Horned and uncastrated animals will fetch much less money, if you can find a market at all, so we have finally had to resort to castrating and dehorning our calves in order to sell them!

Double Suckling

Our cows have their own calf but also adopt a second calf that is bought in from dairy managers who do not want to keep all their calves. We have developed a very successful system which ensures that more calves are raised by mothers in social groups, rather than being confined to calf pens and bucket rearing, and this has proved very worthwhile economically as all our cows have enough milk to raise two good youngsters (Table 26).

This does, however, involve manipulating the behaviour of the mothers: they have to learn to behave nicely to their adoptee, and

Table 26. Economic performance of double suckler beef cattle compared to single-suckled cows and dairy cattle

Figures for 1981 in Britain:		
Incomings	*Feeding costs*	*Margin over concentrates*
16 calves 6–9 months @ £245/calf average 2/cow, per cow = £490	Concentrates @ 45 kg/cow = £25	£490 – £25 = £465
Ministry figures for dairy cows same year:		£280
for single suckler cows:		£50
Figures for 1992:		
Incomings		
£750/cow/year		
(see Table 2 (p. 66) for details)		

they are hoodwinked to believe they have had twins (Kiley-Worthington, 1981 and 1983). Although this does not cause any evidence of distress or suffering, nevertheless it does interfere with their normal behaviour and we would prefer, now, not to do this. Our next ploy therefore is to select animals to twin so that, perhaps in 30 or 40 years, we shall have a herd of pedigree twinning South Devon cattle! We would never use surgical manipulation or foetus swapping or hormones.

The only problem with our suckler cattle is that many of their youngsters are slaughtered when they are the right size for beef. However, our animals are now such tough, hardy, long-living and productive dual-purpose cattle that we are beginning to sell at least the heifers for breeding, and perhaps some of the bulls eventually. There will always remain a few animals that are not good enough who will have to be sold for meat.

Dairy Cattle

Fulfilling the criteria listed in Tables 24 (p. 195) and 25 (p. 198) is often more difficult for dairy cattle because they are kept indoors much of the year in restricted environments, fed high-protein foods, separated from their calves, and often have no courtship or sex as artificial insemination is used on them (see Table 21b, p. 183, which shows the degree of behavioural restriction for typical dairy cows). Diseases are frequent and drugs are used routinely in most dairy herds, particularly antibiotics for mastitis, and hormone treatments to try and combat infertility. They frequently have severe problems with their feet, very short lives in the herd (p. 39 and Table 1), and not infrequently show evidence of behavioural distress (disturbance of circadian rhythms, neuroses concerning milking or coming into the shed, stereotypes and abnormalities such as tongue-rolling, inappropriate standing and lying behaviour, extremely disturbed behaviour when their calves are taken away, excessive maternal behaviour to each other, and so on).

Every commercial milk producer will tell you that there is no other way to produce milk economically. This is not true. For many years we have managed this without a) separating the mother from her calf or b) behaviourally restricting her, and c) we have always had a bull and allowed her sex and courtship.

The dairy cows live with our suckler cattle. In the evening, they come into the shed and the calves are restricted to a calf creep overnight (Figure 17). In the morning the dairy cows are milked and after milking the calves are allowed to run out with their mothers. The advantages here are that we have better, bigger calves which sell for more, we have fewer labour demands since there is only once a day milking (unless there is a very heavy producer when we may have to milk her in the evening as well), the cows run with bulls and conceive easily. We have had no foot problems in 20 years, nor behavioural problems or abnormalities, and they live longer in the herd—some of our cows have been in the herd for over 14 years. Another big advantage is healthier calves with much less demand on labour and possibility of infection. Rearing calves from buckets is a highly skilled and highly risky procedure; cows are really much better at it . . . surprise, surprise. Our dairy cows are fed a small amount of grain (1–4 lbs) when they are milked and the calves have ½–1 lb each in their creep at night. For the rest they have grass, silage, hay and straw, kale and rye (if we are growing it) with the others.

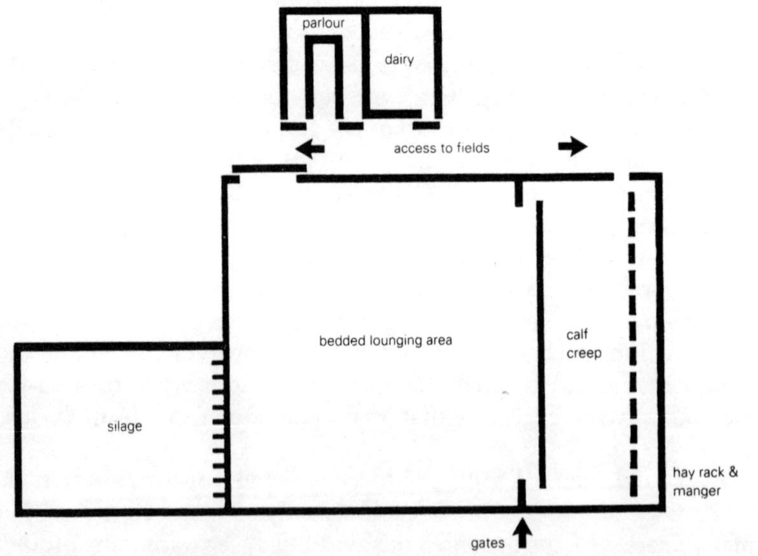

Figure 17. Dairy cow yard with calf creep area. Calves shut into creep for either day or night time. Cows milked once a day.

We take less milk each lactation, but we have more lactations and hence more lifetime milk production from our cows. This is financially a very acceptable way of managing the cows (Table 27).

Of course we have few cows, but there is no reason to believe that this system could not work with a larger herd, although the very large herds (over 100 cows) are inappropriate anyway as they become bigger groups than the cows would naturally form and appear to have difficulty adapting.

The dairy cattle managed in this way are economic as Table 27 shows. In the last ten years more milk has been produced in Europe than could be sold. The solution to this, the government felt, as mentioned in chapter 2, was to give each farmer who was producing milk at that time a quota over which he was not allowed to sell milk. These quotas were handed out just after we left Milton Court and we were not given one. The results were not predicted. In the first place milk production did *not* decline. If the big producers wish to sell more milk than their quota, they buy in someone else's. The small producers are unable to do this because buying quotas is very expensive and they do not have the capital. This is a gross injustice which is strangling change in the dairy industry because there is practically no way into it, and the big farms get bigger and richer.

Table 27. Economics of the dairy cow and calf

Dairy cow		
Production (milked once per day)	1,000 gal/lact = 100 lb cheese @ £1.80/lb	£180
	calf sold at ± 12 m	£450
	Total incoming	£610 per cow
Outgoings: food (home produced) + grass + labour		£250
	Profit	£440
plus milk/cheese etc. for house @ £10 per week		£520 per cow

Sheep

In Britain, and many other parts of the world, most sheep are raised outside and allowed to perform most of the behaviour in their repertoire. They associate in groups, have males with them at least when they would have in the wild situation, raise their own young, and most of the time eat appropriate foods.

As we have seen, they have had some enormous effects on the ecology of large areas of Britain in particular, deforesting in some cases and preventing tree re-establishment by grazing pressure (p. 132). They have had and are having a similar effect in vast areas of Australia.

Before keeping any animal the large-scale environmental consequences must of course be assessed; it is usually only over-population or over-consumption that has such degrading effects. The natural ecosystem can, as we have seen, manage to cope with introduced species without their causing changes unless they become too numerous. In lowland Britain traditionally sheep were fitted into the farming pattern and because of their different grazing habits and catholic tastes they can be a considerable attribute, provided they are properly managed and not too numerous.

It is important to keep sheep in particular out of gardens and woodlands, especially those where young trees are establishing as they will eat the saplings. Cattle, oddly enough, can be slightly less destructive; they simply stamp on them! Fencing therefore has to be good and effective. Cut and laid hedges are of course ideal but it may take some time to establish them. Meanwhile 5–6 strand electric fences can be quite effective and use fewer resources. The wind power can be supplied by a small windmill. The bottom wire of the fence does, however, have to be kept cleared of growth onto it to stop the fence earthing. If there is one place not quite right, you can bet your bottom dollar the sheep will find it and hop over or through with no problem.

Prolonged overstocking with sheep, particularly in the lowlands, results in sheep diseases being particularly rife. They are, however, relatively easy to control with non-invasive or environmentally acceptable biodegradable dips, anthelmintics (worm-killing drugs) and vaccines, provided the animals are treated at the right time of year. Rotational grazing and proper management of grass also helps dramatically. When we brought our highland sheep down to Little Ash, they had never had it so good, and they tucked in to grass so plentiful it was hard to believe, by their standards. They became fat, but it was not long before the down side of the lowlands hit them—foot rot—which they had never encountered in the tough highlands, and various other diseases endemic in the lowlands. There is compulsory dipping in Britain to try to eradicate 'scab', a skin disease of sheep, but there is no legislation about the disposal of the dip and there are some particularly environmentally unpleasant dips used. I asked the inspector once what he recommended farmers to do with their old dip water. This, he admitted, was something the government had swept under the carpet.

Modern developments in sheep management are, however, not in line with our criteria. It is now encouraged to bring sheep into barns to lamb. The result is that enormous barns are built to be occupied for a few weeks a year (and scar the landscape all the year). The sheep are intensively housed in these barns. When they lamb, unless the ewes are kept in individual pens (300 ewes each in individual pens take a lot of feeding and watering, fencing and generally looking after) there is very frequent mismothering . . . the lambs and ewes get muddled up before they have mutually imprinted on each other. During this time the sheep are usually fed concentrate food manufactured and grown somewhere else.

The amount of resources, and problems that are daily, in this system are extremely high. One of the major problems arises because the ewes are often helped with lambing (shepherds will stay with them day and night), and helped to suckle their lambs to start with. Thus there is no selection for sheep able to cope and produce good healthy lambs by themselves. Within two generations this can have a marked effect and flocks have more trouble lambing and being maternal.

Many flock masters thus have a phenomenal number of reproductive problems in their sheep (good for the veterinary profession!) which we (who give minimum supervision to our sheep who are out and get on with it themselves) never have. The number of lambs that we sell at the end of the day is no different (189 per cent lambing: each ewe has an average of 1¾ lambs) and certainly they cost a lot less to produce, environmentally, in terms of resources, time and energy and even money. Sheep (like the majority of animals) are extremely good at looking after themselves; it seems ridiculous to manufacture problems by preventing them from doing this . . .

Intensification of sheep at other stages of their growth is also growing. In some university research departments and commercial farms they are developing feed lots similar to those for cattle. As a result sheep are showing more frequent evidence of distress, stereotypes and abnormalities—wool-eating is one of these. These developments do appear absurd. We have a wool stack and a lamb mountain in Europe, and every lamb killed has a subsidy paid on it—one can be paid to eat one's own lamb!

Instead of this approach, if more sheep are really needed, which I doubt, there are many areas they could use and live on which at present are not integrated with animals. Sheep, for example, are very good lawn-mowers and they do not leave heavy footmarks. They could be integrated in the management of golf courses, playing fields, parks and even private gardens with a little thought and appropriate management. A much more acceptable development, surely.

Our sheep are kept out all the year, although we do have an upstairs room in the barn that they can use when they want during the post-Christmas period, prior and during lambing. They do not have their tails docked, and they are not castrated. They run with rams most of the year. We keep them for their wool and their milk, and some are sold as fat organically produced lamb. When we milk the ewes, again, the lambs are not sold off or separated and put into intensive units (the usual fate for lambs of milk sheep). Our ewes are separated from the lambs overnight (the lambs are herded into a creep) and they are milked in the morning. The rest of the time they run with their lambs. We are also breeding some pedigree long wool sheep for milk and we hope to sell more as breeding sheep in the future. The lambs are often shorn before they are sold in the summer so that we can control strike better and have the lambs' wool spun.

It is thus relatively easy to fulfil all the criteria for ecologically and ethologically sound environments with sheep.

Horses

One of the ideas of the farm is to integrate the horses into agricultural production again, and to give them equal consideration with all the other animals. I have published elsewhere the details of how to manage horses in an appropriate way (Kiley-Worthington, 1987b). What is clear is that the current way of keeping horses in single stables is *not* appropriate from the horses' point of view. They have a host of behavioural problems and show very high levels of distress. Reducing behavioural restriction, developing ecologically and ethologically sound environments for them can be done very easily and makes life easier for horse and human, and much less expensive. It is not only working horses that can be kept this way: the same horses can compete even at international level. Our horses have won major awards in racing, long distance, dressage, driving and so on.

We keep our horses in family groups, we do not wean the foals, and the stallion runs with the mares. They are fed hay and silage, and grains only when necessary for work and fitness. They have a barn they can go into or out of at will. We also think carefully and do research on how they should be educated. As a result we have no behavioural problems in our horses and have not had for six generations. Compared with the behavioural problems listed in Table 20 (p. 177) for horses of this quality kept in the standard way, we have clearly much more closely fulfilled the guidelines.

The only one of our guidelines that we have not managed to fulfil is not castrating the young males. If we wish to sell them, even though they behave beautifully, people are scared of buying a 'stallion' in Britain. People have incorrect preconceived notions concerning their behaviour, because stallions in Britain in particular are almost always pathological as a result of being isolated and treated inappropriately from early youth. We have tried to sell uncastrated three-year-olds but finally had to have the colts castrated. Now we have them castrated young as this is probably less distressing. Before we can sell them properly to appropriate homes, we shall need to change the horse-owning public's ideas!

Pigs

Pigs are originally forest-dwelling, family-living animals. The sow and litter stay together until the litter mates are mature. If they are boars they then go off to seek their fortune, or are persuaded to do this by the visiting boar. If sows, they will stay around for a while and then go off to build a nest and have their litter. Sows and litters come together and may indeed live together sometimes, even cross-suckling their offspring and taking turns baby-sitting.

Outdoor pigs obviously can usually perform more of the behaviours in their repertoire and may not show distress. On the other hand it is also possible to keep pigs in straw yards with rooting areas, or in a family pen system (Stolba, 1982).

Our few pigs are allowed access to woodland for part of the year, and other parts of the farm from time to time. They travel around in a 'pig caravan' which is moved with the seasons. They dig up the garden in the winter, glean the corn in the late summer, go to the woodland for the acorns, and some of the time are restricted to areas around the building. They are kept in family groups. The sows are permanent residents, but the young pigs are sold or killed when they are up to pork or bacon weight. Keeping a boar is expensive. It does not pay unless one has at least ten sows, and we do not have waste products of that quantity to feed them (kitchen waste, whey and corn sweepings, for example). Finding a boar is one of the major problems for the small pig producer. There are many restrictions on the movement of pigs, but it still is usually possible, with determination, to either borrow a boar or take the sows to stay with a boar for a while.

All the other criteria are fulfilled in our pig management. They can be kept away from shrubs and gardens, and from other areas they have a habit of digging up, by electric fences. They are easy to teach and can become delightful companions. It is a great pity we can't find other roles for them except for meat. We are working at it—truffle digging, perhaps . . . if we had truffles.

Poultry

The chickens are also scavengers, and with careful management they can be kept around the building picking up fallen grains, grains from the cattle faeces and so on. Large numbers cannot be kept in this way, however. There is much research at the moment continuing with alternative housing and management for hens because of the public pressure concerning broiler and intensive battery systems. Some good practical ideas are coming out of this work, but it is still difficult for the chickens' environment to be able to fulfil all the criteria we have outlined. Our backyard hens do, and they are remarkably efficient in terms of producing eggs and meat for little input.

Ducks we find more sympathetic. I suppose it is just that they tend not to wreck the garden which after months of work is something difficult to forgive! They need water, and we have a multi-species eco-pond where they swim and dabble about. They can be good layers and the spare males are sold or eaten. They are good for slug control too, and we often encourage them into the garden at particular times of the year.

Geese are grazers and can often be fitted into the garden system, keeping the lawns down and using pieces not often used. Their eggs are lovely if one has a large appetite, and again the young geese are the ones that we sell or eat.

Turkeys I have mentioned before. Free range turkeys for Christmas have a very ready market and they will be no more destructive than chickens and can eat the same things. We have tried breeding turkeys—the bronze and the black are the only ones who can breed nowadays. They are amusing to have around, puffing and blowing with outrage and desire.

It is relatively easy to fulfil all these criteria for the poultry provided one does not have thousands of them, but in an ecological farm one would not have anyway. They can also be profitable. The problem may be foxes, but this does not mean one kills the foxes, only that one shuts the poultry up at night and ensures that during the day the fox does not come. This can be successful with the dogs, humans or just placing of poultry houses.

Other Animals

There are many other animals that can be fitted into the farm system. For example, at the moment we are particularly interested in fibre-producing animals, angora rabbits and llamas (see p. 165). Dogs are crucial workers on the farm for us. Other animals that could be part of ecological farms in different parts of the world include elephants, deer, antelope, peccaries, tapirs, yaks, musk ox, buffaloes and so on. There are many animals that farmers might decide to raise according to where they live. The important thing is that the animals and their management fulfil the vast majority of the criteria listed in Tables 18 (p. 173), 24 (p. 195) and 25 (p. 198). If this cannot be done in many respects, then that animal husbandry should not be practised anyway, and certainly not on an ecological farm. It may of course be difficult to fulfil all the criteria, as we have seen, but with thought and determination it is possible to get very near with all species.

If this is done, then perhaps we shall be one step nearer living in some sort of symbiosis with other sentient beings, and indeed in a position to learn from them as well as them learning from us. Mutual development, perhaps, towards greater understanding of each other but also of the world we live in. It could be said that this might be as important, if not more important and constructive an area of epistemology (study of knowledge), as just the human approach!

10 Women, Agriculture and Food Production

When we reflect on the behaviour of animals, women or men, their similarities and differences, we find three main features for deliberation:

1. Are the differences in the way women and men are treated and think about their abilities in agriculture the result of culture, and how does culture affect sex roles in agriculture?
2. Are there genetic sexual differences which would preclude one sex from being able, for example, to work as a food-producing farmer, and if so what are these?
3. To what degree is the past experience of the individual important in what she can and cannot do, or thinks she can and cannot do? This must consider both genetic and cultural effects, but also those personal experiences which may mould, and may even change, the way the individual assesses herself and her abilities.

1 Culture

For some time now in western paternalistic societies (Capra, 1984) women have traditionally not been recognised as particularly important in the production of food from the land. It is a 'man's job' in which women are not expected to concern themselves beyond giving their husband a hand with the books, the secretarial work, the dairy, the poultry and picking the vegetables. For the rest, their prime responsibilities are preparing and serving the food, and clothing and looking after the children. Although the choice of foodstuffs to present to the family rests primarily with the women, the money with which to procure them is not their responsibility.

They are also not expected to be solely or even primarily responsible for decision-making regarding their children's upbringing.

Their accepted role is really that of housekeeper and cook (Capra, 1984).

The fact that they often do have other responsibilities does not negate this statement; it remains true that as a general rule they are traditionally not expected to. Anything more is the result of volunteering and/or pressurising, or being pressurised by their family or husband. Thus they are not key workers in the system. If hubby is ill it goes without question that a man must be hired to cope with the farm or small-holding. Some young women *may* be hired in agriculture, but they become 'honorary men' while they are working. The wife and mother is very rarely one of these. She is considered not to have time or the ability to be able to do anything particularly consistent on the farm, what with the demands of house, maternity and child-minding. This cultural norm is in large part maintained and reinforced by the women themselves, for many reasons, but often because they may never have known anything different and believe it to be the way it must be.

There have been times in the West when women had equal opportunity on farms, such as during the Second World War when land girls were employed to replace the men who had gone to fight the war. Curiously, after the war, when they had proved that women can be as effective on the farm as men, there was strong pressure to get them off again as the men returned from fighting. They did not always have the option of staying employed in agriculture, and the country reverted to its previous attitude to rural women as being incapable of competing with males for jobs in farming.

It is only in the last decade or two that attitudes to women in agriculture have changed in law. In fact, it is still rare to find an employer who does not select men for most jobs on the farm. A woman has to be very obviously superior to any male applicants if she is to be employed in any job requiring some physical work. It is still women who are employed almost exclusively in the traditional female roles in agriculture, of housekeeper, farm secretary and office worker.

Since the western anthropologists were indoctrinated with the western cultural idea, it was the role of the inferior housekeeping maternity-machine women which was considered by them as normal *homo sapiens* behaviour, and thus the proper and normal way of organising society. So it became a general rule, and all societies were interpreted through this perception: hunter-gatherers, agriculturalists and pasturalists. *Man* became the hunter and provider, woman the maternal machine and home-minder and cooker of the food provided by the man. Both academic and popular books explained everything, from the need to acquire land to swollen breasts and copulation throughout the month, on the bases of *man* as hunter and provider of food (e.g. Ardrey, 1963; Morris, 1967; Mead, 1956).

I had spent my professional life being trained as an ecologist and ethologist and trying to untangle a little of how animals organise their societies, and I had not come across this type of thinking until I was asked to teach some anthropology. In part this was my own oversight, and perhaps also I had early become conditioned not to think seriously about sexual roles. Along that path, for me, could easily lie exclusively domesticity, cooking, cleaning and taking orders; it had happened to some of my closest, brightest and most interesting friends. I had none of these skills and absolutely no wish to develop them.

These assumptions, made by both male and female anthropological writers, seemed to me quite extraordinary. The lack of evidence for their statements made them even more ludicrous: how could they be taken seriously? Was anthropology about the pursuit of knowledge or about wearing a certain shade of spectacles?

Where I had grown up in Kenya, among the Kikuyu, the food production and agriculture was the exclusive domain of the women, who also, of course, prepared and served the food, and bore and raised the children. This was no freak quirk of particular economic accident, it was fundamental to their culture, and as I subsequently found out, of other agricultural societies in Africa. These responsibilities were no 'rights' that were fought for; there was no need to demonstrate their equal ability with the men in agricultural work (as still remains the case in Europe, the United States and Australasia today); their superior status in this regard was expected by men. The men, meanwhile, had their own areas of responsibility, which mainly revolved around making 'important decisions' about wars, justice and marriages, gambling games and buying old motor cars . . . much like men all over the world!

Thus although the women in such societies were bought and sold and appeared to have an inferior status and few rights, nevertheless they were, and are, the main contributors to the survival of the individual and the society because of their role as child- and food-producers.

In the 1970s, as a result of the feminist movement, both academic and popular anthropologists began to question some of the assumptions concerning sex roles in non-western societies, and found that in many that appeared male-dominated, such as Asian societies, African agriculturalists and Aborigine hunter-gatherers, it was usually the women who were the providers of the staple foods (Morgan, 1972). Soon it was demonstrated that certain physical characteristics indicated that humans were not and never had been mainly carnivorous. The men provided some food, generally meat and animal protein, but this was usually for feasts, high days and holidays, and did not feed the family every day. The women, often with the help of small children, grew the staple foods, or collected them and controlled and tended the small livestock. Only among the nomadic pasturalists, it seems, where the bulk of the food derives from animals, did the men play a fundamental role in providing food for the family.

The colonial powers, and subsequent United Nations efforts, until very recently concentrated on agricultural education for the men in all countries. Through this book we have argued that the approach and training of agriculturalists has often been ill-advised and has had unfortunate consequences in many parts of the world (see chapters 1 and 2).

However, much of this agricultural education had remarkably little effect in 'developing' countries. This appears largely to be due to the fact that it was geared to the men, who were considered the dominant members of the society and therefore in charge of agriculture. The women, unsullied by western 'education', have continued to grow and provide the food for the family in the traditional ways (and to bring up their children with their own cultural background). In retrospect this now seems a very lucky mistake of the West, a silver lining to the agricultural cloud.

We now know that if large-scale changes are to be made in peasant food first agriculture, then it is to the *women* that the ideas and education must be addressed. But it has been argued through this book that it is *not* large-scale changes in agriculture that are needed. Where these have occurred, they have undermined and radically changed many societies. The participants have been converted from self-sustaining, independent peasants to members of the consumer society. This has been very good for business in the West, particularly for the multinational corporations, but has it added to the total sum of human happiness, and how have women fared in this change? It is doubtful that it has contributed to the happiness of humans—the recent famines bear witness to this. As to the role of woman, as her control over food production is eroded, so too is her security, self-respect and control over her own and her children's survival and destiny.

Is it too late to reassess what is appropriate agriculture in most of the 'developing' countries? Can this retain the social order, a woman's happiness and self-respect, and even increase her security and independence? Have we in the West, as a result of the massive expenditures on agriculture and agricultural research, discovered *anything* useful for her, to help her towards these ends?

Perhaps it is a consideration of appropriate constructive suggestions, stemming from the increase in knowledge in basic subjects such as ecology, ecological agriculture, etiology (cause) of diseases, soil science and swopping of local agricultural technologies, which will result in helping the traditional women peasant agriculturalists to produce a little more with no more energy, in a self-sustainable and practical way. Inevitably there will be some who farm better than others, techniques that will succeed better for some than for others; there will be few golden rules for how it should be done.

2 **Genetic Sex Differences**

Darwin (1859) argued that only the 'fittest' individuals in societies survived, and Wilson (1975) applied this thinking to the behaviour of individuals. He, and others, suggested that the behaviour of those who leave most offspring will be selected and be the most successful and common in society.

In order to leave as many surviving offspring (which themselves reproduce) as possible, male mammals have two possible strategies: they can mate with one female and then invest time and energy in ensuring the survival of her offspring, or they can invest in mating with as many females as possible rather than spending this scarce time and energy on the subsequent survival of a few offspring. Females, however, particularly mammals, have more risk with reproduction and invest much more time and energy in bearing each infant (they have prolonged pregnancies and prolonged periods after the infant's birth when it is unable to survive without being fed and often protected by them). Each infant thus 'costs' more, and the number they can produce is relatively small compared to that which a male could produce during his lifetime.

Thus, genetically speaking, it is more important to the mammalian female that her offspring, in which she has invested considerable time and energy, should survive, than it is to the male, since he will have invested less time and energy per infant (copulation and the production of sperm are physiologically less demanding than pregnancy and parturition). Thus it is that the majority of male mammals tend to be promiscuous, and try to father as many offspring as possible, whereas females tend to have powerful physiological and behavioural mechanisms to ensure that they are strongly bonded to their infants immediately after birth, and that the infant which has been very 'costly' to them survives.

There are many different cultural veneers to these basic biological factors, but normal females, without exception, are genetically, physiologically and behaviourally programmed to invest greatly in caring for their infants after birth. In particular this involves feeding them. In many societies (particularly western society until the last three decades or so) women had lost their self-reliance since they had often become dependent on monetary income (usually from a male) to supply the wherewithal to provide food. However, in rural societies this need not be the case. Here the women are able to fulfil their intense need to provide for their children by growing the basic foods themselves. In many areas of the world the resources to do this are relatively small: a patch of land 20m by 20m, some seeds and a digging implement.

Thus I suggest that females have a basic biological 'need' to provide for their children, and an inability to do this because of cultural or physical demands may cause feelings of insecurity, and unwelcome dependence. Only by being in control of the food production for their own children can women retain their self-reliance and self-determination.

However, there are of course biological differences between male and female humans, let us not pretend otherwise! One of them is that men can as a general rule run faster than females, and another that they can, again as a general rule, lift heavier weights and are stronger than females. There is little reason, however, why these facts should preclude women from being just as effective as men at food-growing. Neither the tilling of the soil nor the handling of large animals requires brute strength—if it does, then the job is not being done properly. It is only through ignorance, or when agriculture becomes a money-making industry, that implements become so large and heavy, and animals so numerous as to be unhandled and untrained, that brute force may become an asset in agriculture. One way which anthropologists like to use to distinguish humans from other mammals is that humans commonly invent and use tools. Do men have a monopoly of ingenuity and invention of appropriate technology, or can women figure too?

It is also possible that there are fundamental differences in the functioning of the brain in males and females. For example, it has been suggested that females have a different search image and find it easier to switch attention and perform several tasks at once, whereas males tend to be more persistent in attention. The degree to which these differences are cultural rather than physiological or genetic is, however, presently disputed.

Cultural sex differences in agriculture

In the twenty years we have been training male and female students in ecological agriculture (more than 100 to date) to grow food and look after animals, it has become strikingly obvious that females in particular are inhibited by their culture. They consider that they are unable to do certain tasks, so they don't try. Those who overcome this cultural overlay and are determined and prepared to learn are no less proficient than the men, so there appears to be no physiological difference between the sexes in their ability to grow food!

Only too often, alas, women in western agriculture at least have only themselves to blame. They want to have their cake and eat it. So they want equal pay and opportunity with men in agriculture, but at the same time they consider it acceptable, because they are women, to plead lack of strength or energy for a task, menstrual pains, or that they must leave to live with their boyfriend. They can be extremely irresponsible towards their jobs, and will drop the job for a relationship or a small family problem. With such a reputation they are unlikely to be considered as competent, responsible and capable as men in agriculture, where the seasons and animals don't wait.

I have to admit, on the basis of experience, as a female and an employer, that I tend to favour employing men for both agricultural and research work, not because they necessarily have more ability or skills, but because, as a result of their cultural background, men usually tend to be more responsible towards the job and are less likely to leave, throwing up the whole thing (and our project). In general this is because society expects it of them. It does not help the women's cause to plead women's problems: this only results in self-fulfilling prophecies. If women are to reverse and change this cultural belief and receive equal treatment in skills heavily dominated by men, then they must, just like the first female doctors and lawyers, be seen to be more responsible, and better at the job, than the men. The harvest must go on, and the animals still have to be fed and milked, even though the herdswoman has menstrual pains!

One task in both food first agriculture and conventional industrial agriculture that women are in general said to be better at than men is looking after young animals, be they chicks, piglets, calves, lambs or foals. This is hardly surprising in view of women's biological role of child-bearing and minding; perhaps women are also more able to empathise with the young or sick animal or human, and therefore are better at nursing. In some cases, however, this can be over-used by females: we have had female students who could not see the wood for the trees.

As a general rule, again, women are perhaps better than men at attending to detail, although in the process they may lose sight of the whole. In food-growing it may well be this attention to detail that will produce the crop, not the grand overall plan which the men might go for more readily.

There is no doubt that the physical costs of child-bearing and minding, and of food-growing and preparing, can be excessive for peasant women, particularly when they have many children. Their lives may indeed be short, brutish, exhausting and relatively unrewarding. However, with birth control and co-operation between women there is little need for this in the last quarter of the twentieth century, in any culture.

It is perhaps in cultures where women are forbidden by religion or tradition to use birth control that there is now the most need for women's movements and pressure groups. There is little doubt that human population increase is one of the two greatest problems confronting humans, and this will continue to grow (Erlich, Erlich and Holden, 1977).

It seems absurd that religious or cultural dogma, invented and upheld by male-dominated groups, should not only have the power to contribute on a massive scale to increasing this problem, in Africa and South America for example, but that it should also cause great suffering to millions of individual women and children: many women stuck in such cultural blind alleys are worn out before their time, condemned, as they grow older and more tired, to watch their endless stream of children die from malnutrition and disease.

3 Individual Experiences

There are, then, obvious biological differences between the sexes, and possibly some behavioural ones too, but there is no evidence to suggest that any of these should exclude women from producing food, or even detract from their ability to do this. If anything, their biological role of producing and caring for their infants predisposes them to be keen to provide for their own infants' and other animals' needs. The prime one of these is of course food.

So how can the individual woman contribute to the happiness, security and self-worth of herself and her children? This may well vary from culture to culture, but it may help her to know that there is no reason for her to doubt her own abilities in these spheres, whatever her cultural background and past experiences. But there are costs. For example, we have found that the only thing the female and male students *really* have to learn is that, as my mum used to say when I was unable to undo a screw or lift a chair and had dissolved into a frustrated rage: 'There is no such word as can't.'

If you stand around waiting for someone (usually a man) to fix your car, lift a chair, or plough the field, then soon you will forget this little maxim, and it will no longer be true.

Where there is a will there is always a way; if you can't do it, it is because you don't want to, not because it is impossible. There is no reason why you should always want to do it at all, of course, but why not honestly admit it, rather than presuppose that it can't be done by you or perhaps anyone?

Another thing that has become abundantly clear to us over the years is that the attitude must be one of 'making do', of understanding that the proverb of the self-sustaining farmer is not 'if a job is worth doing, it's worth doing properly', because that way you starve, but rather 'it must be done somehow'. Then there should be no stopping you. Don't be ashamed of 'bodging'; at least you have *done* it, which your critic probably has not. However, don't forget to go back and improve it when you can!

Based on our own work, even in a difficult climate where soils are reputed to be poor and the area generally described as agriculturally marginal, once an adult (female or male) is trained and efficient she should take no more than two hours a day through the year to provide her own food and that for two children. She then has many other hours to laugh and sing, to talk to her children, to earn money, to read and study, to do all the other things she might like to do. Then the world is her oyster, and we are back where we were when we were all hunter-gatherers! Even the Bushmen or the Aborigines, living in very difficult unproductive areas of the world, only take about two hours a day to gather their food. The rest of the time they have for leisure. Work was not invented until the origin of agriculture. One wonders why we ever bothered!

To do this, however, women must examine how they spend their time, and use appropriate technology to ensure, for example, that they do not have to spend half the day walking to the well and back to collect water, which is then not enough to water any plants, only to drink and wash a little. Digging is another occupation that can take an enormous amount of energy; and it is energy, *not* money, that counts when it comes to surviving. Washing clothes can be a time-consuming, repetitive activity on which women in developing countries spend a great deal of time and energy. Collecting and preparing food, particularly grinding corn, cassava or maize meal, can also use up much time, and in many countries with deforestation, what takes most time of all is collecting firewood. It is easy to use cow or camel faeces to cook the meal, but as we saw in chapter 1, this encourages the down-grading of the ecology and will make matters worse for the next generation, if not the present one.

A combination of understanding a woman's day and of putting heads together to try to reduce the work most wasteful of time and energy—which the women could well do without—is surely what is needed to help in the future. For example, at the present time providing electricity and washing machines for peasant women in tropical areas is far too complicated, requires too much expertise, and takes too long to install and maintain. However, it would be a great improvement to provide water nearer home, to do the washing in and possibly to water some of the vegetables and flowers around her hut, so long as it did not require a lot of technology and maintenance. I have already extolled the virtues of the ram pump (p. 156) which we have discovered since moving to Devon. There are many millions of places in Africa where it would be an extremely useful gadget. It could be installed, made and run by the women without any great engineering background, and it works for years without any other fuel. If women like to meet at the well and chat, why lay on running water into every house, why not just have the ram pump lifting the water to a tank nearer the village?

Digging and weeding is something most of us women peasants could do without, or at least reduce the time we have to spend on it. One possibility there, again in Africa, is to harness the readily available, usually idle donkeys and train them to pull little ploughs and cultivators which can be made from wood by local women for local women, in just the same way as the basket-making skills they already have can be adapted to make the harness for the donkeys. The donkeys are there, although the women may need some help in training the first few. Again it is a question of putting our heads together and swopping information, local technologies, skills and ideas to make all our lives easier.

Firewood is another important aspect. Here growing and planting of trees, planting of hedges around gardens and looking after them (with water, perhaps, from the ram pump) will, even in five years in Africa if the right trees and shrubs are planted, make an enormous difference. Again, mutual help and exchange of ideas from different cultures with different technologies and traditions can be very useful.

Then there are some things that in tropical countries do not use a fraction of the woman's time and energy that they do in temperate climates. In cold countries, for example, children must be kept clothed and warm, and so dirty more clothes which are difficult to wash (water must be heated) and difficult to dry (no hot sun much of the year). Because they wear plenty of clothes to keep warm, the young children also have to wear nappies, requiring more equipment and washing. In tropical climates young children do not need to wear nappies or have many clothes, and the clothes will dry fast in the sun.

In many so-called developing countries there are extended families and communities. The small children always stay with the mother while she works in the field or house, travelling on her back. When the child is too large for that, then child-minding is no problem because of the extended family and co-operative sharing of the women, and the assumption that as soon as he or she is capable, a child will help in one way or another, by watching or playing with his young sister, tending the sheep and goats, and so on. This is something the western-type countries could well learn. In Europe or the United States, if a mother wants to work on her allotment, go shopping or just have an afternoon away, it is a veritable problem. She must have equipment such as prams, trolleys, play-pens, toys, motor cars or public transport, and must either hire other women to look after her child for the time, or organise an elaborate time-tabled swopping of children between other nuclear consumer women and their children. If she was working at food production, she would have to pay almost all the food she grew on having her child looked after to enable her to do it.

These are just a few ideas to show how we could help each other, apart from the delight and increasing richness that such a detailed exchange and development of individual friendships between cultures would bring to our lives. As a result of thinking about all these things, we now have a plan. This is to ask all interested women in particular, but men too, to contribute to the air fare and maintenance of some African, Indian and other Asian peasant women to come to our farm for a while to *exchange* information, learn what they wish and teach us what they wish. While these women are with us, any contributor may come to visit us, and get to know and exchange information and cultural ideas with the woman she has helped to bring here. When the women go back to their countries, then the contributors, when they take their holidays or go on safari to Africa, can go and see them, help on their farms if both wish, but primarily make and retain contacts and friendships between cultures. Perhaps their children could exchange schools for a while. How else are we going to build internationalism and escape from arms deals and political wrangling? Any interested readers should contact 'Women's Eco-Agriculture Exchange' through the publisher. There are already women from Ghana, Kenya and Zimbabwe who would like to come.

Although women in some parts of the world have certain cultural barriers to overcome in order to be independent food-producers for their family, be self-reliant and self-respecting, they are genetically, physically and psychologically able to do this; indeed, Food First Ecological Agriculture, needing as it does a holistic and 'caring' approach to the biological world, might well sit better on their shoulders than on those of men.

The field is wide open, sisters: over to you.

11 Ecological Agriculture for Africa, Japan, and Australia too?

Even in 1971 the problems facing agriculture did not look easy to solve, but as luck would have it, a year after moving into Milton Court Farm and beginning to mull over the directions and possibilities, we were offered jobs in South Africa for a while, in my case to research and teach wild animal behaviour at the University of Pretoria, an Afrikaans university. Living in the 'bush' studying eland and blesbok suited me well, and allowed me time to learn about and look at agricultural development in Africa. Frequent trips to Swaziland, Lesotho and Botswana emphasised the problems and the urgent need for an agricultural 'development' rethink.

However, the South African power structure of the time and I did not get on too well. I did not make any political noises or statements, but I did try to teach my Afrikaans students to think and question. One bright cold dawn Rebecca, the woman whom we employed to do the housework in our tin *bondoki*, shook me and my husband awake: there were ten plain clothes white policemen in our peasant farmyard. My husband's visa had expired the night before and they had come, they explained, to take him to the airport for deportation. He had been born a South African but had taken British citizenship and therefore they could not arrest either him or me and hold us without trial as they could any South African citizen, but they could kick us out of the country. Luckily, unknown to them, he had renewed his visa the night before. About a month later, I arrived in my laboratory one morning to find a note on my desk: my appointment was at an end and I must go within a month, leaving half-finished projects and students with no supervision.

It so happened that we had visited Swaziland a weekend or two before, and found that there was a new Nuffield project to teach schoolchildren to grow vegetables at school with the idea that they would then go home, teach their parents and grow and sell vegetables. They would then be able to pay their school fees, and eventually the middle class and expatriot Swazis might not have to buy so many of their vegetables from South Africa. They were desperate for volunteers to go and live at the bush schools, start the gardens and teach the children.

After nine months in White South Africa, this project seemed exciting, both professionally and socially. My two boys and I therefore drove off to our new bush school: Ekukhanyani in Swaziland. My sons, aged six and eight, were to attend the school. Both of them had been at whites only schools in the rich suburbs of Pretoria, and the change in life-style and social contacts was dramatic for them. Sam, the elder one, had 50 children aged from 8 to 23 in his class. The children took years out sometimes in order to help at home, or to earn money. They had never seen or touched a white child and leaned over in class to touch his hair and wonder at how straight it was! This drove Sam wild with annoyance, but young mammals, including humans, are adaptable, and he soon developed an involved social life, and found that there were some bright students in his class even though he had the language advantage. The one thing they all had in common was enthusiasm to learn. Pip's class was in the infants' school and the teaching medium was Swazi. There were 80 in his class! Inevitably it was not easy for either of them, but the sun shone and people laughed, as they do in Africa.

I taught the higher classes. Although in some respects the course had been very well thought out, it had been designed by middle-class European academics who had no real experience of what it was to be a Swazi outside the consumer society. Like all of us, they were the products of their culture, with a profound belief that progress, education and civilisation were what *they* had experienced in their own culture. Quite quickly major problems became obvious.

For example, the majority of land in Swaziland at that time was available for common grazing for the local people's cattle, sheep, donkeys and goats. In order to start a garden around the school, therefore, it was necessary to fence off the area. How and with what was the fencing to be constructed? The Nuffield organisers had provided the school with fence posts, wire, nails, staples, and tools, and immediately the project become one that was done at school but would be impossible at home. How could any peasant outside or on the edge of the consumer economy possibly afford posts, wire, nails and so on? They could not even rub enough money together to pay school fees. The course organisers had assumed that somehow or other such money would eventually be available and, in particular, that the job *could not* be done without it, despite the fact that the local people were adept at making stockproof boundaries out of local materials. It would have been possible to plant and maintain hedges out of appropriate local plants, including unpalatable and prickly plants. Hedge-making, cutting and laying is not a technique that is universally understood in many countries except for parts of rural England. It might be of very considerable importance in many other parts of the world, as well as in Africa.

The organisers also provided a diesel pump to pump water for irrigation. It was assumed, again, that without this and without irrigation the project could not work. Even at the school, never mind at the children's homes, there was little likelihood that anyone would know how, or even be motivated, to maintain the diesel generator. Africa is littered with broken-down internal combustion engines of one sort or another that have been donated by well-meaning philanthropists—tractors, pumps, cars, generators. Not everyone is a mechanic or wants to be, and how can you continue to buy fuel and spare parts if you have no money? Is money the answer anyway? Despite every effort to impose a universal consumer economy, there still are people who live happy, healthy lives without it!

Two questions emerge from this. First, was it necessary to irrigate at all? Could one not have concentrated on crops and vegetables that, by planting at the right time and in the right microclimate, might be able to live and grow without it? Secondly, if one did want to grow other crops that required irrigation, was there no other way of pumping water using what was already there, or what the local people could make and maintain? Draught animals and humans can work simple pumps. It might have been possible, with help initially, for the local people to make and maintain ram pumps (back to them again!) which are very simple and require no energy but that of running water (see p. 156).

Here I was confronted practically with the limits in thinking of even the thoughtful philanthropists and professional educators. It emphasised the even wider range of problems and considerations that must be included in our 'appropriate alternative agricultural strategy'.

So how can we help the peasant women of Africa? We can help them with an attitude of mind, encourage them to question, to try new things and develop further their existing skills; we can swop information. They are very much better in many ways at using the equipment they have than we are in the West. As a result of research, we should have more knowledge of soil structure and how to test it. Use and construction of microclimates might be useful, and planting and managing of woodland, propagating trees so that they do not have to go great distances for their firewood. We can help and swap information in many ways—a sample has been given on pp. 225–6.

POSSIBILITIES FOR ECOLOGICAL AGRICULTURE IN JAPAN AND AUSTRALIA

Recently I had the opportunity of visiting these two countries, as different as chalk from cheese, yet oddly enough, in some ways confronting similar problems. I was on a study tour, considering animal welfare and looking at the problems of agriculture and possible solutions. Could Ecological Agriculture help confront some of their major problems, or are they only the result, in Japan of very high human populations, in Australia of a difficult climate?

Japan

Japan has a human population estimated at 123,116,000 in 1989. It is a highly industrial nation and relies on imports for much of its food. This may well be acceptable economically at the present time and enhances trade pacts between Japan and food-producing nations. However, what are the long-term environmental effects of this? And can Japan's human population go on increasing indefinitely? We have already shown in chapter 1 that the consumer society, based on materialism and monetary wealth, cannot provide the answers to the world's problems. In the future, with a growing scarcity of resources, there is no way that it can go on. There are just not enough materials to allow the population of the world, even at the present, to live as profligately and irresponsibly with resources as the average citizen does in the United States,

Australia, Europe and Japan, let alone the increasing human populations of the future. The Club of Rome came to this conclusion in 1970 (Meadows *et al.*, 1972). But what is being done about alternatives? If in the future (as we hope will happen, to ensure our children's futures), environmental concerns become central to governments, countries like Japan may well be affected dramatically on several fronts:

1. The markets for their goods will be cut as their trading partners will not have the money or the priorities to buy items not crucial to their survival or environmental improvement.
2. The raw materials, most of which are imported to Japan for manufacture, will become difficult to obtain as countries use their own resources to manufacture goods.
3. As human populations and self-sustainability of countries grow, present food-exporting countries will consume more of their own food, and not grow it at environmental costs primarily for export.
4. Since Japan is reliant on food imports, she is extremely vulnerable in the event of natural or human disasters. A lesson could be learnt from the situation during the last war, when populations generally were lower.

Thus there is a case to be made for Japan to produce more food from her own resources for her human and other animal populations, without causing further environmental degradation. But how can Ecological Agriculture help?

First, a brief and very incomplete assessment of Japanese food production as I saw and heard about it. The human populations are highly and successfully urbanised. They live almost exclusively in cities around the coasts on the flatter land, the remainder of the country being forests which are managed for timber, conservation and wildlife reserves, with relatively small areas for golf courses, ski resorts and so on. However, very few people indeed live in rural environments away from the cities or in the forests, and they do not seem to want to.

The farming is done usually by part-time farmers who have very small holdings (less than 1 to 5–10 hectares). They specialise in growing rice, which the government subsidises, and as a result Japan is self-supporting in rice, their main staple. They also grow vegetables, herbs and fruit. They often have between one and ten or so cattle who are usually kept in small stables or yards and zero-grazed (food is cut and brought in to them, or they are fed imported food). Much of the cattle food, including such bulk low-quality feeds as hay and straw, is imported from the United States. There are also large-size intensive chicken and pig units, although I did not see them. Again, the vast majority if not all of the food to feed these animals is imported. Even bulky fodders such as hay and straw are imported from the United States and thus very expensive. There are of course some exceptions to this type of husbandry, but at present it is the norm. The farmers use high levels of inputs of all types in order to try and maximise their returns, and as a result there are all the usual problems confronting agriculture elsewhere in the world (see chapter 2), only perhaps more acutely.

The Japanese are extremely fussy about their food: they will only buy food that is very beautifully presented and they take enormous trouble and skill to make and present food well. On the other hand, they do not yet seem to have followed the western city dwellers' serious concern with *how* the food has been produced and whether or not there are likely to be serious health risks in eating it due to pesticide residues, and other chemical treatments. My feeling, however, is that, because of their deep concern with food and its importance both physiologically and symbolically, it will not be long before this becomes a prime concern of the Japanese consumer.

With the enormous growth of living standards in Japan in the last decade or two, they, like other rich people worldwide, eat a great deal of protein—not only more than they need physiologically, but more than is good for them in the long run. Thus the usual diseases of over-eating high-quality foods are relatively common, although their cultural preoccupation with aesthetic appreciation—they are masters of 'good taste'—seems to have controlled at least the number of fat people. Coming from Europe, one cannot but be impressed by the slimness and well dressed appearance of the general Japanese public, and also their phenomenally good manners.

Their wealth and change in philosophy as a result of close contact with American consumerism has dramatically changed their food habits in the last thirty years, and increased their consumer demand for other products. This is illustrated by their demand for fish and seafood of all types, and like every other nation in the world, they do not manage and control their fishing in as self-sustaining a way as they could. The environmental results of this are:

1 The Japanese commercial fishers travel worldwide to try and harvest more fish. This often results in water feuds and water boundary disputes, and in over-fishing of fish stocks. The Japanese whaling activity is only one aspect that has recently hit the headlines.
2 They are prepared to buy seafoods for high prices from other nations. This encourages these nations, particularly poor countries in South Asia, to over-fish and exploit their own seas.

The same story is repeated with the supply of many other foodstuffs to Japan, and other biological products whose exploitation causes environmental degeneration (for example, the exploitation for sale by Australia of huge areas of the indigenous old Kauri forests of the south-west for pulping, without any effort to manage it sustainably).

At home, like other industrial nations, they have some major environmental problems that they must overcome, such as air and water pollution from industry and industrialised agriculture. These problems are in many ways more acute in Japan than in many other industrial nations because of the higher population and few other resources, and because they have concentrated so successfully on following the industrial path.

Concern for animal welfare has not as yet hit the public eye in Japan in quite the same way as it has in Europe and the United States. Certainly, among scientists involved in agricultural animal husbandry, there is great interest in many of these ideas, but since there is confusion worldwide on these issues, at the moment they are at the stage of assessing the different approaches and controversies of the West.

Although, in Japan, legislation on farm animal welfare lags behind that in Europe to some extent, I was not convinced from what I saw that it is behind in practice. Everywhere in the world there is enormous need for improvement and clarification of ideas on these issues. I was, however, very impressed with the open-minded and interested approach of Japanese scientists in this area. In many countries, although philosophers may argue the toss, the scientist and veterinarian think they know the answers and are not prepared to discuss them further. I did not find this to be the case in Japan.

There is no reason why the ideas and practice of animal husbandry outlined in chapters 8 and 9 should not be equally applicable in Japan, aided by one very important factor that may well have considerable effect on consumers and producers in the future. Japanese society until around thirty years ago was predominantly Buddhist. This meant that they ate no meat and relatively little fish. Although probably the majority in Japan now consider themselves irreligious, I find it difficult to believe that in such a short time such a strong cultural background can be shrugged off. Indeed, evidence that this is not the case is the ever-increasing concern of the Japanese with their own cultural background, the preservation of numerous shrines and places of religious and cultural interest, and the enormous number of people who visit these. The vivacity of the Japanese culture and its success is illustrated by the remarkably low crime rates and little violence (by western standards) in their large, congested urban populations.

Conversation with scientists, thinkers, intellectuals and ordinary, everyday people convinced me that the Japanese are very aware of their materialistic attitude to the world at present, and not particularly satisfied with it. Although it has brought riches and much comfort, I sensed a wistfulness in their manner when the subject was bought up. It will surely therefore only need the further reconstitution of the Japanese cultural belief system for animal welfare and environmental concerns to become highly important and popular. My feeling is that in Japan there is much hope that such changes will take place in the not so distant future, particularly when they realise that the West, far from having the solutions, is thrashing around in a quagmire of contradictions. Indeed, Japan may soon be setting the pace for western countries where the problems are not necessarily so severe and the people are well encased in what is in some ways a less sympathetic cultural belief system.

What are the answers? Could Japan become more self-sustaining in food and other biological resources, and if so how? She could of course become entirely self-sustaining in food, but this might not be necessary or desirable provided she only trades for food and other products which have been *sustainably* produced. First of all the Japanese will have to revert in part to their old cultural beliefs and at least cut down the amount of fish and meat they eat. As a conditional vegetarian in Japan (I only eat meat and fish if I am convinced it has been raised in ecologically and ethologically sound environments) it was sometimes difficult for my hosts to feed me because practically all their dishes have some meat or fish in them. I had many beautifully cut and presented pieces of aubergine in batter! In traditional restaurants every dish has symbolic meaning and is a work of art, and every other dish is vegetarian. It would therefore not be too difficult for them to turn their considerable ingenuity and innovation to producing vegetarian dishes and to make such food sought after by the trend-setting youth. There is a tendency in this direction in the West.

Ecological Agriculture would clearly have a very important role to play in Japan, producing greater net production, diversification, environmentally sound, high-quality foods which would be sold often direct to the consumer. The recycling of human sewage would be of particular importance here, since the number of animals raised for food would be reduced. They already seem to have a general idea of conservation (preservation with utilisation) of wild and semi-wild resources, and encourage consumers to use Japanese materials rather than being dependent on imports. Why, when so much of Japan is covered in forest, do they buy the Australian trees? They must surely be able to produce sufficient timber for their own use, even if not for export. They may also have to change their forestry strategy so as to provide the woods required.

There are farmers and even councils in Japan who are interested in ideas like these. I went to see a particularly interesting village near Myasaki—Aya, where the council encourages everyone to farm without putting chemicals on the land, and local business people run a shop selling only organic and locally produced food and crafts. They also give grants to establish craftsmen and cottage industry using local materials. I visited a bamboo basket-maker who was in great demand as a teacher, as well as making baskets himself, and a small farm producing free range eggs from chickens that lived under the mulberry trees kept for the silk worms; there was also a small factory spinning and dyeing (from local herbal dyes) the resulting silk.

Such developments are certainly needed in Japan, and there is great interest and hope in the possibilities.

Australia

I have to admit to suffering a considerable culture shock on flying in to Australia after a stimulating and exciting month in Japan. Bussing through parts of Queensland and then New South Wales as I made my way from one university to another, I saw only too well how important it was that the Australians should change their attitudes and stop exploiting their phenomenally large, productive and exciting land. Environmentally it is as much a disaster as the southern Sahara. I felt as disturbed and miserable, looking out at the millions of purposely ring-barked dead eucalyptus trees dotting the plains and hills as far as the eye could see, as I had in parts of the Sudan and Northern Africa or the Scottish Highlands. Environmental exploitation at its worse! Moreover, these semi-tropical and tropical areas cannot stand the same level of abuse as at least temperate Scotland can. In Africa, however, the denudation occurs largely because people are struggling to survive, it is often life or death; in Australia the motivation is profit. The people here have not been threatened with death by starvation, but they would not otherwise have had such a good short-term monetary gain.

At the same time as seeing the advance of the semi-desert upon the species-rich, self-sustaining indigenous forests (fractions of which have been 'preserved' as nature reserves and so are there for comparison), I also became aware of the phenomenal ease of living and the rich resource reservoirs of Australia. Practically nowhere else in the world has it so good today—they have more land, mineral resources, potential renewable power supplies, wood, fruit and practically every other natural resource per head of population than any other country in the world. Capping it all is the very equable climate. It may be hot and suffer droughts or dry seasons, but don't forget that in temperate lands we have a five-to-nine-month winter *every year*, not one year in five or so, when little or nothing will grow (even if we water it), and it is also much more complicated to live comfortably. We have to have heavy, warm clothes, washing and drying facilities for them and heat to keep warm. In the winter three-quarters of the time it is dark, and there is no time of the day or night when it is comfortable to work outside without protection . . . unlike the farming regions of Australia.

So what went wrong? After considering this at length and talking to many people, I wonder if the reason is that the vast majority of white Australians, both rural and urban, charming, informal and delightful though they are, remain fundamentally urban in attitude.

Many of the early settlers and convicts were from highly urban backgrounds, with the result that they took with them their urban philosophy of alienation and fear of the natural environment when they struck out into the Australian outback.

They had to outdo, overcome and tame wild nature by any means available, despite the fact that by, for example, African standards there were fewer difficulties to contend with—no lions to eat you or your stock, no elephants to smash your crops, no bilharzia or malaria or any other endemic debilitating disease. There was plenty of room, and almost everywhere (originally) wood and stone for building and an abundance of firewood. You could build your house from materials right on your own land, and use your own natural resources to make fences and hedges. There was no winter; although there might be a slightly colder season or a dry one, it was not in the same category of difficulty as a European winter. There was plenty of land for the taking, and on top of all this the sun shone almost every day. Even water was available: wells could be dug and sometimes the water came up of its own accord! Ram pumps could have been installed in the rivers for irrigation (if there had been any rurally-based folk from the West of England among them). Earth dams could easily be made, even without a constant running water supply, and they held water without the treatments necessary for dewponds on the chalk. By peasant standards it was a paradise.

But most of the white settlers were not peasants or even rurally-based people. They were quite unfamiliar with how to live in a rural environment and develop a sustainable life-style. They had no background of living with nature (or if they had they forgot it), of having appropriate local rural technologies to cope with it, of developing sustainable ways of doing this for the years ahead; no tradition of muddling along and fitting in with the natural ups and downs of the living world. For most Australians, nature was merely there to be exploited in order that they might make money and eventually, God willing, be able to get back to the city in Europe where they really wanted to live. From the start industrial agriculture became the main trend. Indeed, many of the farmers, having built their farmhouses from imported corrugated iron (goodness knows how much this cost and how many animals and humans perished trying to get it to the far-flung parts of Australia, but it became an essential ingredient of Australian 'rural' life) and cut down all the trees around them, found that their home had become so ugly, hot and unpleasant that they drove off in the 'ute' (large four-wheel drive vehicle) to hire a house in the nearest town, and commuted daily to their farm to work! They still do this, and when one sees how thoroughly unaesthetic their farmhouses can be, one can understand their reasons. Why did they not, and still not, think for themselves a little more? Some of them did, of course, and developed lovely homesteads which others would visit and appreciate, but did not then apply to their own way of life.

Why did they not see what their farming strategies were doing to the land, how the dead trees exposed the soil, cut down the circulation of nutrients and water, and in a few years reduced the grass growth that they wanted so badly? That by profligate use of the readily available water the water table became more saline until they could only grow certain crops? On one of my earlier trips to Australia, some twelve years ago, I went to lecture at a well-known agricultural college in Queensland. They were interested in the ideas of self-sustaining agriculture and the technologies for achieving this. Among other things we discussed the increasing soil and water salinity problems which have now escalated even further. Afterwards our host took us to see his farm. We found him pumping out water from the aquafer at a very high rate to irrigate his lucerne. I asked him why, and his answer was that lucerne was relatively saline tolerant and was the only crop he could grow in the area . . . so he continued to pump the water out and seemed to think this was a sensible thing to do! This type of thinking is only too common, of course, not only in Australia but throughout the world.

Bring in people with no peasant/rural background technologies or familiarity with living in close contact with the natural world, but rather with a well-founded belief in money as the panacea for all ills, give them land but little if any appropriate understanding, and you have a recipe for land disaster. Their treatment of the indigenous population, including extermination and destruction of their culture, may well have resulted in part from seeing how easily they were able to survive and enjoy their lives while these white urban people were finding rural life very, very difficult. Almost ever Australian classic literary work, of which there is a rising tide in the wake of their two hundredth anniversary, tells the same tale—how very hard life was—yet the Aborigines apparently survived and prospered before the whites arrived! Communication lines were cut, however, and now it is too late to learn from them. Much of their technology and environmental expertise has been lost by separating the children from their parents to teach them to be 'good Christians', and by raising them in urban environments.

The Australian rural Judaeo-Christian subculture then arose out of this background. It involved demonstrating how tough and hard life was in the Australian bush, how tough and hard the breadwinner, dominant family member, must be, and how he must fight and overcome nature (or any competitors, including plants and animals) if he was to be considered a good and responsible person. In the traditional Judaeo-Christian belief system women were to be protected but were not granted similar status to males. This was not unlike the culture of the western states of America but, philosophically, *very* different from the indigenous populations of both continents.

Thus the lack of rural peasant philosophy and technologies left the new Australians in a difficult position. Life was and is hard for them. The result was that much effort and expense were devoted to developing agriculture through research institutes, advisers, teachers, colleges, and universities. There was no need for discussion of previous experience, attitudes and experiences as there was in Europe in particular, because few of these had been considered positive or helpful. It was not necessary to wait for the old class, with their old attitudes and technologies, to die and their sons to be educated in the modern agricultural system before things would change: all were set on the industrialisation of agriculture. It was and still is possible to keep in touch with the most up-to-date developments in Europe and the United States, through constant swopping of 'experts' and 'advisers'.

By the early 1970s, however, things were looking bad. More land was turning saline, more water resources were being exhausted or becoming saline, more topsoil had gone; the desert was expanding and the wild forests were almost extinct. A few perceptive people began to think of alternatives. One of these is the

Tasmanian Bill Morrison, the originator of Permaculture (see pp. 56–7), who published his first book when we were working along similar lines developing Ecological Agriculture.

Now the 'in-word' in Australia is 'Sustainable Agriculture', and almost every university and college is hiring people to teach this, offering and teaching classes in this form of agriculture (see p. 56). This may be a step in the right direction but it is also worrying. Sustainable agriculture is interpreted predominantly in terms of planting tree windbreaks and small farm woodlands (Campbell, 1991), certainly an important first step, but there are many more changes that will have to take place in Australians' attitudes to farming and food production before their environment is likely to be sustainable, never mind self-sustaining.

I am full of hope that this can happen. In Western Australia, which is, I found recently, more exciting both environmentally and in terms of human development than the patriotic east, I came across a group at Murdoch University thinking and working along these lines.

I was impressed with the incredible resilience of the natural vegetation and fauna of Australia, its ability to survive and thrive after fires, droughts, floods and human interference causing denudation of soils. As in the Western Highlands of Scotland (see chapter 6) the problems are not biological—give the natural ecosystem a chance and it will recover—but human attitudes, thinking and lack of knowledge of appropriate technologies.

There are a few farmers thinking this way now in Australia, the first stage being to begin an ecological upgrading of the farming system rather than continuing the biological exploitation. Previous chapters indicate how this could be done. In particular:

1 Predicting and catering for a hard time of year by conserving animal and human food, and keeping only the number of animals that can be supported in this way.
2 Recycling wastes instead of causing pollution problems as a result of agricultural run-off (for example, in the Mandurah river of Western Australia).
3 Thinking about multi-species grazing, including fitting in indigenous animals such as wallabies, kangaroos and emus. In fact better grazing management is badly needed and would be quite practical in many areas.
4 A less profligate use of resources should be encouraged,

particularly water (which in urban and rural environments seemed to have no rationing or control), and renewable energy production, especially from solar power.

5 Use of indigenous materials for building, and probably planning controls even in rural environments, to try to fit buildings and their design into the environment, are crucial. During one month in Australia we only came across one building in all the places we visited that had been constructed out of the natural mud bricks and vegetation. Even within the National Parks tin roofs rule, although there is no reason why mud bricks, local sustainably managed wood, and local thatching materials should not be much more widely used.

6 Fire risk is real and must be catered for, but should not dominate the design.

7 Creation and use of micro-climates has not been seriously taken on board and should be incorporated in the system.

8 Animal management systems need to be very seriously reconsidered.

Intensive pig and poultry husbandry systems seem particularly absurd in this environment. These were developed in areas where land was scarce, rainfall high (giving rise to mud) and temperatures very low in winter. In much of Australia there is plenty of land, well-draining soil with relatively low rainfall, and no very low temperatures. One cannot help thinking that these are ideal conditions for the development of outdoor, ecologically raised pigs and poultry, for which there is a growing market, in Europe in particular, if more are produced than are needed in the area. They could have special crops grown for them, which they could then root up themselves, could live in family groups and be rotated around gleaning and digging. The intensive piggeries I saw in Australia not only required large amounts of resources and fuel to build, feed the animals and keep going, but also, because they had enclosed the pigs and the temperature rose, even in the spring in Western Australia, they required air-conditioning to prevent them over-heating! Outside the temperature was a pleasant spring day! Intensive husbandry systems of any animal would seem particularly misplaced in Australia, but at present grants are given to encourage them!

Sheep and cattle management is usually ethically more acceptable. The animals run out on large areas and can perform most of the behaviours in their repertoire. Three areas, however, need very serious investigating:

1 *Winter/dry season feeding*
Often the animals are left to die rather than maintaining the appropriate number.

2 *Use of surgery*
Particularly 'mulesing', whereby the rumps of sheep are cut open in order to heal with scar tissue which will then be less prone to strike (flies lay their eggs, maggots hatch and in appropriate moist wool they feed on the skin and flesh of the sheep). The number of animals that die from this surgery is unknown, the number that suffer are all the sheep that are mulesed. The argument given in its favour is that without mulesing the animals would be more vulnerable to strike and would suffer even more. There are, however, other ways of overcoming strike, for example by dipping and by use of biodegradable back sprays. The reason why these are not used much in Australia is because the farmer has so many sheep and such large distances over which to gather them that it does not pay. If this is so one cannot help asking why one man should have so many sheep and so much land, and one should also look at what other work he does. Has he really not got the time? Perhaps it is rather that he does not *want* to do this, and the cost of losing a few from mulesing is not so important!

3 *Transport*
Cattle and sheep are transported enormous distances live to be sold or fattened. Of itself transport may not necessarily cause suffering, but since it may take four to five days to transport cattle from the north to the south of Western Australia, for example, and they have never been enclosed before, it is something that needs very careful investigation to see how it can be done without causing massive suffering.

Sheep are not only transported live across Australia in trucks, they are also put live on ships, for export to be killed in Muslim countries, where it is the practice not to pre-stun the animal before killing it. This surely must be outlawed. The animals die in their thousands, having already endured difficult and trying truck journeys. One observer mentioned to me that few people bathed near the loading beaches because there were usually dozens of floating sheep carcasses that had been thrown in. Is this really acceptable, for the sake of a few extra dollars? The sheep could be slaughtered on the farm or nearby, and transported as carcasses.

The last area of concern is the general attitude of Australian Greens and Conservationists towards conservation. There is of course a case for 'preservation' of particularly interesting or last-resort eco-types all over Australia, but that this attitude should extend to all National Parks and all indigenous vegetation is misguided and in the long run will not benefit the conservation of much of the Australian fauna and flora. Again it originates with the urban attitude of the Australians, and there are groups thinking this way now in Europe, too (see p. 149 for further discussion). Such urban-thinking people, who have a growing following in urban environment, have no experience or understanding of living with Nature in a reasonably symbiotic way, where each is used by and uses the other and so can stay around, let us hope, indefinitely. It is assumed that man is somehow not part of Nature, but is separate and unique in some way, not controlled by the same rules, and that humans will inevitably always have a destructive influence on the Great Outdoors, the Wild Yonder.

The urban rural-based Aussie will find it difficult to move away from his homestead without his Eski (a large insulated box in which he carries bags of ice and cold beers), his 'ute', his sun block lotion (to prevent the sun giving him skin cancer), his hat and dark glasses (in case he gets sunstroke), his tents, swag (large canvas bag with plenty of clothes, mattress and sleeping bags), billy (pot in which to boil water), several four-gallon demijohns of water, dried and processed food, and so on and so on. This for a weekend break in the bush for a bush family. Is this necessary for rural survival, or even comfort—unless one has had no other experience? Contrast this with the light load of the Bushmen or Aborigines, even when living in the desert, or a Masai, Bedouin or travelling tribesman from anywhere in Africa . . .

The result of this predominant attitude in Australian Conservation is that the National Parks, which after all are supported by the rate-payers, the public, are usually, but not always, areas where humans are forbidden to do various things. They are not allowed to bring their dogs in for a walk, ride their horses, canoe, camp, fish, swim or in any other way enjoy and participate in the environment except by walking, picnicking (no control over what the humans eat in the National Parks) and making fires (a traditional Aussie cultural norm that even the National Parks have not dared outlaw).

Why should it not be possible to educate the public who go there to behave responsibly and enjoy and be part of the living system, just like the birds, goannas and insects? Little effort is made to do this in many National Parks (one exception is a forest gorge near Armidale in New South Wales, where information and encouragement to do all these things with discretion showed unusual foresight and perception). Rather it is assumed that everyone will behave irresponsibly, throw their beer cans around, catch all the fish, break the trees, gallop around without regard for anything else. The result is that the rules are not obeyed. The public are irritated, but at the same time they are not given any education on how and where to do things. They pay for these parks, but even on the busiest day of the year a park will have a few hundred visitors at most, and often less than five a day, compared to the thousands a day in some European and United States parks.

Such attitudes do not bode well for the future of the society or the National Parks in terms of their harmonious relationship with the natural world.

The most viable solution is integrated utilisation of these wild areas in a sustainable way, without exploitation and preservation: recognition that we too are part of the natural world, and so are our dogs, horses, cattle and other sentient and non-sentient life forms—whether we like it or not!

My abiding impressions for the future of ecological agriculture and the environment in Australia were positive, again because of the phenomenal resilience of the natural fauna and flora, their adaptability, beauty and profusion. What a biologically beautiful and exciting continent! There is every chance, given a cultural reassessment, that all could go well and build, particularly in Western Australia, an exciting multi-cultural, environmentally sound self-sustaining population of humans and all the others. So, laughing kookaburras (whether or not you come from Western Australia—the white people didn't originate there either), curious chattering pink and grey galahs, who will always come to investigate the bipedal, equally chattering humans, phosphorescent 'twenty-eight' green parrots, good natured people and delightful grey mares, among other residents . . . the future is in your hands.

12 Conclusions

What, then, is the solution to the problems outlined in chapters 1 and 2, and how should we think about agricultural development in the next century?

I have argued that conventional high-input farming is not biologically efficient and creates a myriad of biological, social, aesthetic, ethical, economic and political problems (chapter 2). To continue along this path, in which the agricultural establishment and business generally have a vested interest, *may* make the rich richer, but clearly it will also make the poor poorer and will increase the numbers of humans and other animals that starve (chapter 1). Although I believe that those who developed and are currently involved in this type of agriculture may be well-intentioned, nevertheless they have been misled by their cultural conditioning, the belief in the 'technical fix', inappropriate education and partial and woolly thinking.

There is no evidence that over-production in one part of the world, as the result of this conventional high-input modern farming, helps provide for those starving elsewhere. In other words, food distribution worldwide is not the solution and is not improving. Even if it were to work, and we could one day look towards a Garden of Eden where food produced in one place could be sent to the other side of the world to be consumed, is this desirable? I think not. The inevitable result of such an approach is the control of one part of the world by the other, because one controls the other's food source.

Reduced political independence and eroded self-reliance might be the solution if one society was convinced that it had all the answers, and that God was definitely on its side. But then, this has usually been the belief to date, from the Crusaders to the Vietnam war!

I and others are sceptical of this approach, not only because it clearly has not worked, but because of our delight in diversity of cultures, religions, habitats, political opinions, moral obligations . . . freedom for all living things. One can produce ponderous and valid arguments concerning the environmental importance of diversities and self-determination; but if I am honest it is because of an epistemological quest to find out what the world is about. The more approaches to it that there are in practice, the more we can learn, and the more exciting it is. The more we can question, the more answers we may find.

If, then, a food first agriculture can be dreamed up and can work in providing the operators and other living, sentient beings with a self-reliant life of quality, and which does not have undesirable spin-offs for the entire environment, then this may be one step in this direction. Ecological Agriculture was defined and its tenets outlined in order to try and do this (chapters 3 and 4). Although some farms are run to some degree along these lines, nevertheless there are grey areas, and it was necessary to see if it really would work in, firstly, a relatively easy environment (chapter 5), and, secondly, in a much more difficult marginal environment (chapter 6). The results of these two farms have, I hope the reader will agree, indicated that there is certainly a future in such an approach. They have also pointed out the problems, which have not always been in the expected places, and the areas where more research could be particularly useful. The third farm (chapter 7) further develops some of these ideas.

There is now a galloping interest in what is euphemistically called 'Organic Agriculture'—whatever this means (see pp. 51–3). Not an issue of the main farming magazines comes out without at least one article on how a big-time farmer is reforming his ways and 'going organic'. Membership of the Organic Farming groups is rising exponentially; supermarkets are searching to buy 'organically produced foods'. Politically ambitious young and not so young people are becoming spearheads of the 'Organic Movement' as advisers and media people.

But surely one should not knock this. Anything that encourages people to consider environmental issues and food production should be welcome.

I am not so sure, I am worried about the recent popularity of the Organic Movement, the jumping on the bandwagon by establishment institutes such as research centres, universities and colleges of agriculture, and large profit-orientated farmers. How is it that all their, as they believe, 'watertight arguments' against such an approach (which I have encountered, and countered again and again, from international conference delegates to the small farmer down the road; from the lecturer in a local college of agriculture to the director of a prestigious research institute) have so suddenly been forgotten under pressure from the relatively uninformed public? One is tempted to consider that they never really understood their own arguments, and that they may misunderstand or misrepresent those for 'Organics'. Is not this euphoria just another way of making a quick buck?

As a general rule I think this is exactly what it is. Those who will be teaching 'Organic' agriculture, who will be practising it and advising on it in the next decade, with very few exceptions will not have changed their agricultural paradigm. They will not have examined the problems and their causes and possible solutions . . . they will simply be applying the approach they know, the technological fix, albeit a slightly different technological fix, and jumping on the bandwagon that is likely to attract students, money and expansion. I don't really see things changing much. It takes a strong person to stand up and admit that the world view he or she has been living by and teaching for the last few decades is fundamentally flawed. We have to wait for all those currently educated in conventional agricultural thinking to retire before the next generation will have a real chance of being educated to an understanding of what Ecological Agriculture is about.

There are, however, two related areas in which more genuine and constructive change is likely to take place.

The last decade and a half has seen the re-awakening of a new consciousness concerning other sentient beings, mammals and birds in particular. It may well be in this area that there will be revolution. We have argued that, when in doubt, it is rational to consider that other mammals are more similar to than different from us humans, *until or unless we have evidence to the contrary* (chapter 8). This shifting of the burden of truth has not as yet been taken very seriously. What it means inevitably is that if we have no evidence that animals *do not* think, are not self-aware and so on, then we must assume that they *do* and *are*; and if that is the case, we must draw more and more parallels between how we as humans think and feel, and how they do. When we do this, the floodgates will be opened to an examination of how we treat and use them. The same questions must be raised and answered as when human slavery was abolished.

The other area of hope for genuine development of thinking and acting is feminism. The more women in the West who are prepared to pick up the cudgel of self-reliance and enter the professional world of science, the more chance there will be for a reassessment of what science is or should be, and whether its masculine, manipulative approach, to which we have all been subjected, is the only way in which we can scientifically tackle the problems and understand the answers. For this reason, it may be that science, and in particular agricultural and biological sciences, will progress in the future along quite different but exciting lines.

Thinking, and working practically, for the last twenty years on the ins and outs of Ecological Agriculture has brought great moments as well as disastrous ones, and in retrospect I think and hope it has all been worthwhile. There are no grant-giving authorities, trusts or financiers of any type to acknowledge . . . although often I could have wished for them; the farms, the animals, and the human inhabitants have themselves financed the research. This book is as much to thank them as, hopefully, to entertain, encourage, outrage, or inform others.

Appendix Suggestions for Research

Not needed: Further 'surveys' of 'organic' farms.

Throughout this experiment it has become evident that there are many areas where research is required in order to further the development of a system of more environmentally acceptable agriculture. Until recently, practically all research relevant to ecological agriculture has been done outside the conventional agricultural system (Hodges, 1982). The following list outlines the more important problem areas where organised research is required. In producing this list of research needs it is not intended to suggest that some work has not been, or is not being, done in various specific areas.

1 Farmyard Manure

As the main source of fertiliser in ecological agriculture, a greater understanding of the preparation and use of FYM is needed. Specific topics are: Methods of treatment (composting, etc.) in order to maximise nutrient return and reduce leaching loss of nutrients; time of application of FYM in relation to length of growing season, temperature, rainfall, etc.; optimal applications of FYM, particularly in regard to the effects of different dosages on different crops.

2 Weed Control

The design and use of appropriate mechanical weed controls for field and garden crops, including tractor, horse and oxen drawn machines. The timing of weeding for maximum effect on various crops.

3 Grass and Grazing Management

The multi-species grazing programme (proposed in Kiley, 1974) has shown much promise, with carrying capacities continually increasing. However, further work is required on food selection and intake of different grazing animals. For example, the total intake in a grazing situation of horses and geese is not known, and the details of what plants, stages of growth, etc., are selected by different species, sexes and age groups are poorly understood.

Although work has begun on the concentration of certain nutrients and trace elements by different plant species and parts (e.g. Holder, 1978) this work needs further extension and development in order to be able to devise heterogeneous leys designed to cater for animal selection and to retain nutrients and trace elements within the system.

Quantitative studies on the reduction of parasites as a result of multi-species grazing, and the effect of harrowing on the spread of infective larvae and sward growth, are also needed.

4 Animal Husbandry, Applied Ethology and the Ethics of Animal Welfare

The present approach has been a radical one since it is maintained that animals should be able to perform practically all aspects of their behavioural repertoires, and that these are behavioural *needs*. Thus excessive interference with breeding (e.g. use of artificial insemination, or the curtailment of courtship, or the manipulation of animals with drugs during reproduction) has not been acceptable. Research and practice on the techniques of achieving these aims has already begun (Kiley, 1976; Kiley-Worthington and de la Plain, 1983). Such an approach requires extensive research on alternative housing systems, feeding systems, etc., coupled with proper ethological studies, whilst bearing in mind economic factors.

The ethics of animal welfare and animal rights are an area of great development and discussion in philosophy (see, for example, Clark, 1976; Singer, 1976; Miller and Williams, 1983). Some agriculturalists are unaware of, or do not understand, the arguments which are becoming of considerable public concern. Further research of these issues is badly needed, as also with the ethics of man's responsibility towards nature (Passmore, 1974).

5 Energy Budgets

Although some energy budgets have been given in this paper, the difficulties and probable inaccuracies arising from this approach have not been disregarded. Many more figures are needed for comparisons with conventional units and with natural ecosystems. Much can be learnt from ecologists with respect to methodology, although there are many areas on farms with which they have not been concerned.

6 Appropriate Technology

There is a need for the designing, making and testing of appropriate hand tools and machines for use within a low-capitalised agricultural system; also the making of tools from materials on the farm. Animal-drawn equipment also needs further development using modern materials and materials available locally. Some work has been done on ox-drawn equipment, and the author is presently involved in designing and producing lighter and more efficient horse-drawn equipment; however, much remains to be done.

Alternative energy generators for the farm system are another facet that needs more research and development.

7 Further Development of the Ecological Farm

Since farming is essentially pragmatic and its success or failure to a large extent depends on the efficiency and practicality of the personnel involved, it is essential that further experimental/demonstration farms be set up to test further the ecological farming system in both the developed and the developing worlds. Such farms must be run within the normal economic constraints of the area.

8 Education

At present there are no government supported farms in Britain where students can learn both the theory and practice of this type of farming. There are a few farms which do take students and train them for certificates in this type of agriculture, but they are financed by private funds. There is an increasing demand for places on these farms and it would seem to be long overdue that appropriate courses be run within the national educational structure. In the last part of the twentieth century the interrelationships between the different parts of the biosphere and between different scientific disciplines are becoming more widely recognised. If humanity is to solve some of the large-scale problems confronting the world (see Erlich *et al.*, 1977) it will be essential to have a more 'holistic' approach. It would seem a necessary educational development, therefore, that agricultural research workers, among other scientists, should begin to be trained in the ethical, biological, sociological and political arguments associated with food production.

Bibliography

ADAM, F. H. (1978). The Wad Ramli Co-operative—a Sudanese case study. *Beiträge zur Tropischen Landwirtschaft und Veterinärmedizin*, **16**, 1.

ADSHEAD, N. (1986). *A comparative energy analysis of an ecological farm and conventional farms on the Isle of Mull, Scotland.* Liverpool Polytechnic, Geography dissertation.

AHMED, O. M. M. and ADAM, S. E. I. (1979). Toxicity of *Jatropha curcas* in sheep and goats. *Research in Veterinary Science*, **27**, 1.

ALBRIGHT, J. L. (1987). Human/farm animal relationships. In M. W. Fox and L. D. Mickley (eds.), *Advances in Animal Welfare Science 1986/87*. pp. 51–66.

ALI, A. H. (1977). *Agriculture in the Sudan. Selected Bibliography with Abstracts.* Ministry of Agriculture, Food and Natural Resources, Khartoum.

ALI DARAG ALI (1976). Carrying capacity assessment under annual range type. *Sudan Silva*, **3**, 21.

ANONYMOUS (1974). *Alternative Landbouw.* Centrum voor Landbouw, Wageningen.

ANONYMOUS (1979). Turning the Sudan into the Arabs' bread basket. *Business Week*, No. 2603, 46–51.

ARDREY, R. (1963). *The Territorial Imperative.* Collins, London.

ATTFIELD, R. (1983). *The Ethics of Environmental Concern.* Basil Blackwell, Oxford.

BAILEY, M. (1977). Sweet Prospect for Sudan Sugar. *New African Development*, **11**, 1.

BALFOUR, E. B. (1975). *The Living Soil and the Haughley Experiment.* Faber & Faber, London.

BARNET, T. (1977). *The Gezira Scheme. An Illusion of Development.* Frank Cass, London.

BEILHARZ, R. G. (1988). Breeding birds to excel in modern farming systems. *Proceedings Poultry Information Extension*, Gold Coast, Queensland.

Bell, R. H. V. (1970). The use of the herb layer by grazing ungulates in the Serengeti. In A. Watson (ed.), *Animal Population in Relation to their Food Resource*. Blackwell, Oxford, pp. 111–24.

Benson S. and Duffield, M. (1979). Women's work and economic change; The Hausa in Sudan and in Nigeria. *IDS Bulletin*, **10**, 4.

Best, R. H. and Ward, J. T. (1956). *The Garden Controversy*. Studies in Rural Land Use No. 2. Wye College, Univ. of London.

Blaxter, K. (1975). The Energetics of British Agriculture. *Journal of the Science of Food and Agriculture*, **26**, 1055–64.

Blaxter, K. (1976). The use of resources. *Animal Production*, **23**, 267–79.

Blobaum, R. (1975). How China uses organic farming methods. *Organic Grower Farmer*, **22**, 45–9.

Boeringa, R. (ed.) (1980). Alternative methods of agriculture. *Agric. Environm.*, **5**, v–vi and 1–200.

Boerman, F. H. and Likens, G. E. (1970). The nutrient cycles of an ecosystem. *Scientific American*, **223**, 92–101.

Boyles, D. (1980). The Biggest Ditch. *Geo*, **2**, 2.

Brambell, F. W. K. (1965). *Report of the Technical Committee on Welfare of Farm Animals under Intensive Husbandry Systems*. HMSO, London.

British Cattle Breeder's Club Digest, **14**, 1974.

Britton, D. K. and Hill, B. (1975). *Size and Efficiency in Farming*. Saxon House, Farnborough.

Brown, J. (1982). *The Everywhere Landscape*. Wildwood House, London.

Brown, W. L. (1961). Man's insect control programmes. Four case histories. *Psyche*, **6**, 75–109.

Cadiou, P., Mathieu-Gaudiot, F., Lefevre, A., Le Pape, Y. and Oriel, S. (1975). *L'Agriculture Biologique en France, Ecology ou Mythologie?* Univ. of Grenoble Press, Grenoble.

Campbell, A. (1991). *Sustainable Farming*. Lothian Press, Melbourne.

Campbell, W. E. (1975). *Behavioural Problems of Dogs*. Amer. Vet. publ., Santa Barbara.

Capra, F. (1984). *The Turning Point. Science, society and the rising culture*. Flamingo, Fontana, London.

CARSON, R. (1962). *Silent Spring*. Houghton Mifflin, Boston, Massachusetts.

CLARK, S. (1976). *The Moral Status of Animals*. Oxford University Press, Oxford.

CLARKE, C. (1968). *Population Growth and Land Use*. St Martins Press, New York.

CLARKE, K. W. (1963). Stocking rate and sheep cattle interactions. *Wool Technol., Sheep Breed.*, **10**, 27–32.

CLOUDESLEY-THOMPSON, J. L. (1977). Reclamation of the Sahara. *Environmental Conservation*, **4**, 2.

COLE, D. D. and VAIL, D. J. (1980). *Action Research in Abyei: An Approach to the Identification, Testing and Selection of Appropriate Technologies in a Rural Development Context*. Development Discussion Papers. Harvard University, No. 89.

CRAVEN, J. A. and KILKENNY, J. B. (1976). The structure of the British cattle industry. In H. Swan and W. H. Broster (eds.), *Principles of Cattle Production*. Butterworth, London.

CREGIER, S. E. (1987). The psychology and ethics of humane equine treatment. In M. W. Fox and L. D. Mickley (eds.), *Advances in Animal Welfare Science*. Martinus Nijhoff, Boston, pp. 77–88.

CRITCHFIELD, R. (1978). Crocodiles, cattle, and the Jonglei canal. *International Wildlife*, **8**, 4.

CULPEPPER, N. (1959). *Complete Herbal*. Foulsham, London.

CULPIN, S., EVANS, W. M. R. and FRANCIS, A. C. (1964). An experiment on mixed pasture. *Expt. Husb.*, **10**, 29–38.

DAHLBERG, K. A. (1979). *Beyond the Green Revolution*. Plenun Press, New York.

DANZER, R., MORMENDE, P. and HENRY, J. P. (1983) Significance of physiological criteria in assessing animal welfare. In D. Smidt (ed.), *Indicators Relevant to Farm Animal Welfare*. Martinus Nijhoff, Boston.

DARWIN, C. (1859). *On the Origin of Species by means of Natural Selection*. John Murray, London.

DASMANN, R. F., MILTON, P. and FREEMAN, P. H. (1973). *Ecological Principles for Economic Development*. Wiley, New York, pp. 151–63.

DAY, P. R. (1972). Genetic vulnerability of major crops. *Plant Gen. Resource Newsletter*, 27.

DEEVEY, E. S. (1970). Mineral cycles. *Sci. Am.*, **223**, 169–56.

DICKINSON, A. (1980). *Contemporary Animal Learning Theory*. Cambridge University Press.

DIGERNES, T. H. (1979). *Land Use in a Centralised Periphery*. Lecture given at a Nordic Association for Development Geography Seminar. DERAP Paper, Chr. Michelsen Institute, Bergen, No. 84.

DIRAR, H. A. (1975). Studies on the hydrogen peroxide preservation of raw milk. *Sudan Journal of Food Science and Technology*, 7.

DIXON, P. L. and HOLMES, J. C. (1987). *Organic Farming in Scotland*. Edinburgh School of Agriculture.

DORST, J. (1971). *Before Nature Dies*. Penguin, Baltimore, p. 134.

DUMONT, R. and ROSIER, B. (1969). *The Hungry Future*. Methuen, London.

DUNCAN, I. J. H. (1978). Frustration in the fowl. In B. M. B. Freeman and R. F. Gordon (eds.), *Aspects of Poultry Behaviour*. Edinburgh Brit. Poult. Sci., pp. 15–31.

DUNCAN, I. J. H. (1978). The interpretation of preference tests in animal behaviour. *Appl. Anim. Ethol.*, **4**, 197–200.

DUNCAN, I. J. H. and WOOD-GUSH, D. G. M. (1971). Frustration and aggression in the domestic fowl. *Animal Behaviour*, **19**, 500–4.

EGGERT, F. P. (1978). Preliminary results from plot trials to compare the efficiency of several soil management systems. In J. M. Benson and H. Vogtman (eds.), *Towards a Sustainable Agriculture*. Verlag Wirz, Aarau, Switzerland.

EKESBO, I. (1978). Ethics, ethology and animal health in modern Swedish livestock production. In D. W. Folsch (ed.), *The Ethology and Ethics of Farm Animal Production*. E.A.A.P., Vol. 24, Munich, pp. 46–50.

ERLICH, P. and ERLICH, A. (1982). *Extinction*. Victor Gollancz, London.

ERLICH, P. R., ERLICH, A. H. and HOLDEN, J. P. (1977). *Ecoscience*. W. H. Freeman, San Francisco.

ESSLEMONT, R. J. (1984). *Oestrus detection in farm animals*. Proc. E.A.A.P., Vol. 31, Munich.

FERGUSON, W. (1969). The role of social stress in epidemiology. *Brit. Vet. J.*, **125**, 253–4.

FOLSCH, D. (ed.) (1978). *The Ethology and Ethics of Farm Animal Production*. E.A.A.P., Vol. 24, Munich.

FOX, M. W. and MICKLEY, L. D. (eds)(1987). *Advances in Animal Welfare Science 1986–7*. Martinus Nijhoff, Boston.

FRANK, H. and FRANK, M. G. (1982). Comparison of problem-solving performance in six-week-old wolves and dogs. *Anim. Behav.*, **30**, 95–8.

FRANK, H. and FRANK, M. G. (1984). Information processing in wolves and dogs. *Acta Zool.*, Fenn, **171**, 225–8.

FRANK, H., HASSELBACH, L. M. and LITTLETON, D. M. (1987). Socialised vs. unsocialised wolves (*Canis lupus*) in experimental research. In M. W. Fox and L. D. Mickley (eds.), *Advances in Animal Welfare Science*, pp. 33–50.

FREEMAN. D. M. (1971). Stress and the domestic fowl: a physiological appraisal. *World's Poult. Sci. J.*, **27**, 263–75.

FREY, R. G. (1983). *Rights, Killing and Suffering*. Blackwell, Oxford.

GABALY, M. M. El (1977). Problems and effects of irrigation in the Near East region. In E. B. Worthington (ed.), *Arid Land Irrigation in Developing Countries: Environmental Problems and Effects*. Pergamon Press, Oxford.

GAUDET, J. J. (1979). Management of papyrus swamps. Berichte aus dem Fachgebiet Herbologie der Universität Hohenheim, No. 18. In M. E. Beshir and W. Koch (eds.), *Weed Research in Sudan*. Vol. 1 Proceedings of a Symposium.

GEORGE, S. (1976). *How the Other Half Dies. The Real Reasons for World Hunger*. Penguin, Harmondsworth.

GOODALL, J. Van Lawick (1971). *In the Shadow of Man*. Houghton Mifflin, Boston.

GRIFFIN, D. R. (1984). *Animal Thinking*. Harvard Univ. Press, Cambridge, Massachusetts.

GRIFFIN, K. (1974). *Land Concentration and Rural Poverty*. Macmillan, London.

GRIFFIN, K. (1976). *The Political Economy of Agrarian Change*. Cambridge University Press.

HAFEZ, E. S. E. and SCHEIN, M. W. (1962). The behaviour of cattle. In: E. S. E. Hafez (ed.), *The Behaviour of Domestic Animals*. Baillière, Tindall & Cox, London, pp. 247–96.

HAMILTON, D. and BATH, J. G. (1970). Performance of sheep and cattle grazed separately and together. *Exp. Agric. Anim. Husb.*, **10**, 19–26.

HARDISON, W. A., REID, J. T., MARTIN, C. M. and WOOLFOLK, P. G. (1954). Degree of herbage selection by grazing cattle. *J. Dairy Sci.*, **37**, 87–102.

HARRISON, R. (1964). *Animal Machines.* Lyle Stuart, London.
HASKELL, P. T. (1977). Integrated pest control and small farmer crop protection in developing countries. *Outlook Agric.*, **9**, 121–6.
HEARNE, V. (1986). *Adam's Task. Calling animals by name.* Heinemann, London.
HEINRITZ, G. (ed.) (1982). *Problems of Agricultural Development in the Sudan.* Selected papers of a seminar. Editions Herodot, Göttingen.
HEIRD, J. C., BELL, R. W. and BRAZIER, S. G. (1987). Effects of early experience upon adaptiveness of horses. In M. W. Fox and L. D. Mickley (eds.), *Advances in Animal Welfare Science*, pp. 103–10.
HENDRICKS, S. B. (1969). Food from the Land. In N. S. Freeman and N. R. C. Freeman (eds.), *Resources and Man.* W. H. Freeman, National Academy of Sciences, San Francisco, CA. pp. 65–85.
HILL, S. (1981). Soil, food, health and holism: the search for sustainable nourishment. In D. Kuwr (ed.), *New Principles in Food and Agriculture.*
HMSO (1980). *Codes of Practice on Farm Animal Welfare.* HMSO, London.
HMSO (1980). *Conservation on Farms.* HMSO, London.
HODGES, R. D. (1982). Agriculture and horticulture. The need for a more biological approach. *Biological Agriculture and Horticulture*, **1**, 1–13.
HOLDER, J. D. (1978). *Chemical Analysis of Plant Species Relevant to Ecological Agriculture.* Thesis, Univ. Sussex, 32pp, mimeographed.
HORATH, J. (1977). *Food Production and Farm Size. A reconsideration of alternatives.* Halcomb Res. Inst. Working paper. Indianapolis.
HUFFAKER. C. B. (1958). Biological control of weeds with insects. *Annu. Rev. Entomol.*, **4**, 251–76.
HUGHES, B. D. and DUNCAN, I. J. H. (1981). Do animals have behavioural needs? *Applied Animal Ethology*, **7**, 381–93.
HUNGERFORD, F. (1987). *Ethical concepts on ethology.* Delivered at Workshop on Ethics in Veterinary Medicine, Applied Philosophy Soc., Glasgow, March.
HUNSPERGER, R. W. (1963). Comportements affectifs provoqués

par la stimulation électrique du tronc cerebral et du cerveau anterior. *J. Physio. Paris*, **55**, 45–88.

HUNTINGTON, R. (1980). Popular Participation in Sudan: The Abeyi Project. *Rural Development Participation Review*, **2**, 1, 14–18.

IBRAHIM, F. (1978). Anthropogenic causes of desertification in Western Sudan. *Geographical Journal*, **2**, 243–54.

IFOAM (1986). *Proceedings of a Conference on Sustainable Agriculture*. Santa Cruz, California.

INTERNATIONAL THERIOLOGICAL CONGRESS (1982). Proceedings. Helsinki.

JENSEN, P. (1986). Normal and abnormal behaviour of animals. In M. Thelestran and A. Gunnarsson (eds.), *Ethics of Animal Experimentation*. Acta. Physiol. Scand 128, pp. 1–23.

JEREMY, A. C. and CRABBE, J. A. (eds.) 1978. *The Isle of Mull. A Survey of its Flora and Environment*. British Museum (Natural History), London.

JERISON, H. (1973). *Evolution of the Brain and Intelligence*. Academic Press, London.

JEWELL, P. A. and ALDERSON, G. L. H. (1977). Genetic conservation in domestic animals: purpose and actions to preserve rare breeds. *Proc. R. Soc. Arts 125*, pp. 693–710.

JOHNSTON, B. and ALLABY, M. (1977). Agriculture in Britain as a mature industrial society. *I.D.S. Bull.*, Univ. of Sussex, pp. 42–8.

JOHNSTONE-WALLACE, D. B. (1937). The influence of management and plant association on chemical composition of pasture plants. *J. Am. Soc. Agron.*, **29**, 441.

KEEN, B. A. (1946). *Agricultural Development of the Middle East*. HMSO, London.

KHOGALI, M. M. (1979). *Proceedings of the Khartoum Workshop on Arid Lands Management*. United Nations Univ., Tokyo.

KILEY, M. (1974). The behavioural problems of domestic and wild ungulates with references to crowding and grazing. In V. Geist and F. Walter (eds.), *The Behaviour of Ungulates and Its Relation to Management*. I.U.C.N., Morges, Switzerland.

KILEY, M. (1976). Fostering and adoption in beef cattle. *British Cattle Breeders Digest*, **31**, 42–55.

KILEY-WORTHINGTON, M. (1977). *The Behavioural Problems of Farm Animals*. Oriel Press, London.

KILEY-WORTHINGTON, M. (1980). Problems of modern agriculture. *Food Policy*, August 1980, 208–15.

KILEY-WORTHINGTON, M. (1981). Ecological agriculture. What it is and how it works. *Agriculture and Environment*, **6**, 349–81.

KILEY-WORTHINGTON, M. (1983). The behaviour of confined calves raised for veal. Are these animals distressed? *Int. J. Study Anim. Prob.*, **4**, 198–213.

KILEY-WORTHINGTON, M. (1986). *Ecologically, ethically and ethologically sound environments for animals.* Delta Conference, Boston.

KILEY-WORTHINGTON, M. (1987). The Druimghigha project. An ecological farm in a marginal area. In P. Allen and D. Van Drusen (eds.), *Global Perspectives on Agroecology and Sustainable Agricultural Systems*, 273–83. Proceedings of the Sixth Int. IFOAM Conference. Agroecology Program, Univ. of California, Santa Cruz.

KILEY-WORTHINGTON, M. (1987b). *Behaviour of Horses in Relation to Management and Training.* J. A. Allen, London.

KILEY-WORTHINGTON, M. (1990). *Animals in Circuses and Zoos.* Little Eco-Farm Publications, Basildon.

KILEY-WORTHINGTON, M. and de la PLAIN, S. (1983). *The Behaviour of Beef Suckler Cattle (Bos taurus).* Verlag Birkhauser, Basel.

KILEY-WORTHINGTON, M. and RENDLE, C. C. (1984). Ecological Agriculture. A case study of an ecological farm in the South of England. *Biol. Agric. & Hort.*, **2**, 101–33.

KILNER, P. (1975). Breadbasket for the Middle East. *African Development*, **9**, 1.

KISS, J. (1977). *Will Sudan be an Agricultural Power?* Studies of Developing Countries, Institute for World Economics, Hungarian Academy of Science, No. 94.

KLEPPER, R. *et al.* (1977). Economic performance and energy intensiveness on organic and conventional farms in the corn belt. *Am. J. Agric. Econ.*, **59**, 1–12.

KOEPF, H. H., PATTERSON, B. D. and SCHAUMANN, W. (1976). *Biodynamic Agriculture.* Anthroposophical Press, New York.

LEACH, G. (1975). Energy and food production. *Food Policy*, **1**, 62–73.

LEAVER, J. D. (1970). A comparison of grazing systems for dairy herd replacements. *J. Agric. Sci.*, **20**, 265–72.

LEOPOLD, A. (1949). *A Sand County Almanac: And Sketches Here and There.* Oxford University Press, New York.

LOCKERETZ, W., EPPLER, R., COMMONER, B., GERTLER, M., FAST, S. and O'LEARY, D. (1976). *Organic and Conventional Crop Production in the Corn Belt; a Comparison of Economic Performance and Energy Use for Selected Farms*. Report CBNS-AC-7. 42pp. Center for the Biology of Natural Systems, Washington Univ., St Louis.

LOVELOCK, J. E. (1987). *Gaia. A New Look at Life on Earth*. Oxford University Press.

LUESCHER, V. A. and HURNI, J. F. (1987). Contribution to a concept of behavioural abnormality in farm animals under confinement. In M. W. Fox and L. D. Mickley (eds.), *Advances in Animal Welfare Science, 1986/87*, Martinus Nijhoff, Boston, pp. 67–76.

MABEY, R. (1972). *Food for Free*. Collins, London.

MAFF Bulletin No. 36 (1947). *Manure and Fertilisers*. HMSO, London.

MAMOUN, I. E. (1978). *Bibliography of Agricultural and Veterinary Sciences in the Sudan*. National Council for Research, Agricultural Research Council, Khartoum, Ed. 2.

MARGALEF, R. (1968). *Perspectives in Ecological Theory*. Chicago Univ. Press, Chicago, Il.

MARSTRAND, P. and PAVITT, L. L. R. (1974). The agricultural subsystem. In H. D. S. Cole, C. Freeman, M. Jahoda and L. R. Pavitt (eds.), *Thinking about the Future*. Chatto & Windus, London, pp. 56–65.

MATTINGLY, G. E. G. (1974). The Woburn organic manuring experiment. In *Design, crop yields and nutrient balance*. 1964–72. Rothamstead Exp. Rep., **2**, 98–133.

MAYNARD-SMITH, J. (1966). *The Theory of Evolution*. Penguin, Harmondsworth.

MCINTO, H. J. L. and VARNEY, K. E. (1972). Accumulative effects of manure and N on continuous corn and clay soil. In: Growth yields and nutrient uptake of corn. *Agron. J.*, **64**, 374–9.

MCNAB, P. A. (1987). *Mull and Iona*. David & Charles. Newton Abbot, Devon.

MEAD, M. (1956). *New Lives for Old*. Gollancz, London; Morrow, New York.

MEADOWS, D. H., MEADOWS, D. L., RANDER, J. and BEHRENS, W. W. (1972). *The Limits to Growth*. Universe Books, New York.

MELLANBY, K. (1975). *Can Britain Feed Itself?* Merlin, London.
MESAROVIC, M. and PESTREL, E. (1976). *Mankind at the Turning Point*. Hutchinson, London.
MIDGELEY, M. (1983). *Animals and Why They Matter*. Univ. Press, Georgia, Athens.
MILLER, H. B. and WILLIAMS, W. H. (eds) (1983). *Ethics and Animals*. Humana Press, New Jersey.
MINISTRY OF AGRICULTURE, FISHERIES AND FOOD (1971). *Codes of Practice for Husbandry of Farm Livestock*, HMSO, London.
MOHAMED, A. (1978). The projects for the increase of the Nile yield with special reference to Jonglei project. *Water Development and Management. Proceedings of the United Nations Water Conference, Mar del Plata, Argentina, March 1977*. Water Development, Supply and Management Series, Vol. 1, Pergamon Press, New York, Part 4.
MORGAN, E. (1972). *The Descent of Woman*. Souvenir Press, London.
MORRIS, D. (1967). *The Naked Ape*. Cape, London.
MORRISON, B. (1988). *Permaculture. A designers' manual*. Tagari Publ., Tyalum, Australia.
MORTON, A. (1987). Weighing death against pain in people and animals. *Applied Philosophy Workshop on Ethics in Veterinary Medicine*. Glasgow (unpublished).
MUNRO, I. B. (1980). Stress Indicators in Cattle. *S.V.E., Abstract, London Meeting*.
MURDOCH, W. W. (1975). *Environment*. Sinauer Assoc., Sunderland, MA.
MURRAY, W. H. (1970). *The Companion Guide to the West Highlands of Scotland*. Collins, London.
NATURE CONSERVANCY COUNCIL (1990). *Wildlife on Farms*. Pamphlet. HMSO, London.
NAESS, A. (1990). *Ecology, Community and Lifestyle. Outline of Ecosophy*. Oxford University Press.
OBEID, M. (1975). Utilisation of water hyacinth (*Eicchhornia crassipes*)—introduction. In M. Obeid (ed.), *Aquatic Weeds in the Sudan with Special Reference to Water Hyacinth*. Prepared for the Workshop on Aquatic Weed Management and Utilisation, National Council for Research, Sudan, and National Academy of Sciences, USA. Agricultural Research Council, Khartoum.

Odum, E. P. (1971). *Fundamentals of Ecology*. W. B. Saunders, Philadelphia.

Oesterdiekhoff, P. (1979). *Directions, phases, trends and alternatives on agricultural policy*. Research project on possible ways of developing an underdeveloped agriculture like Sudan: possibilities and limitations of an emancipatory development strategy on the political bases of Pan-Arab co-operation. Universität Bremen, No. 6.

Oesterdiekhoff, P. (1980). *The Breadbasket is empty. The Option of the Sudanese Development Policy*. Forschungsberichte, Forschungsprojekt Handlungsspielraume in Unterentwich elten Agrarland Sudan. Universität Bremen, No. 10.

Osman, A. M. and Voigtlander, H. (1980). Some problems of milk production and milk processing in the Democratic Republic of Sudan, *Beiträge zur Tropischen und Subtropischen Landwirtschaft und Veterinärmedizin*, **18**, 1.

Osman, H. E., El Hag, G. A. and Osman, M. M. (1975). Studies on the nutritive value of water hyacinth (*Eichhornia crassipes*) (Mart.) Solms. In M. Obeid (ed.), *Aquatic Weeds in the Sudan with Special Reference to Water Hyacinth*. Prepared for The Workshop on Aquatic Weed Management and Utilisation, National Council for Research, Sudan, and National Academy of Sciences, USA. National Council for Research; Agricultural Research Council, Khartoum.

Passmore, J. (1974). *Man's Responsibility for Nature*. Duckworth, London.

Paton, W. (1984). *Man and Mouse*. Oxford University Press.

Pearce, S. M. (1987). *An Introduction to Animal Cognition*. Lawrence Erlbaum Associates, Hillsdale, New Jersey.

Pearsall, W. H. (1965). *Mountains and Moorlands*. New Naturalist Series. Collins, London.

Peart, J. M. (1962). Increased production from hill pastures. *Scottish Agriculture*, **24**, 10–12.

Perelman, M. (1975). The real and fictitious economies of agriculture and energy. In W. J. Jewell (ed.), *Energy, Agriculture and Waste Management*, pp. 133–9. Ann Arbor Science Publishers, Michigan.

Pollard, E., Hooper, M. D. and Moore, N. W. (1974). *Hedges*. Collins, London.

POLLARD, N. (1981). The Gezira Scheme—a study in failure. *Ecologist*, **11**, 1.

PREBBLE, J. (1963). *The Highland Clearances*. Secker & Warburg, London.

PROCTOR, J., HOOD, A. E. M., FERGUSON, W. S. and LEWIS, A. H. (1950). The close folding of dairy cows. *J. Br. Grassl. Soc.*, **5**, 243–50.

RACKHAM, O. (1976). *Trees and Woodland in the British Landscape*. Dent, London.

RADCLIFFE, D. A. (1965). Grazing in Scotland and upland England. In D. A. Radcliffe (ed.), *Grazing Experiments and the Use of Grazing as a Conservation Tool*. Symposium 2, Monks Wood Experimental Station, England.

RAUHE, T. and KNAPPE, S. (1971). Systems approach to yield formation with special reference to the humus reserves of the soil. *Arch. Bodenfruchtbarkeit Pflanzenprod.*, **15**, 281–8.

REAGAN, T. (1982). *All That Dwell Therein*. Univ. Cal. Press, Berkeley.

REAGAN, T. (1983). Animal rights, human wrongs. In H. B. Miller and W. H. Williams (eds.), *Ethics and Animals*. Humana Press, New Jersey, pp. 19–44.

REICH, C. A. (1971). *The Greening of America*. Northumberland Press, Gateshead.

REINHARDT, V. (1980). *Untersuchungen zum Sozialverhalten des Rindes*. Birhauser Verlag, Basel.

ROLLIN, E. B. (1981). *Animal Rights and Human Morality*. Prometheus, New York.

ROLLIN, E. B. (1983). The legal and moral bases of animal rights. In H. B. Miller and W. H. Williams (eds.), *Ethics and Animals*, Humana Press, New Jersey, pp. 103–20.

ROLLIN, E. B. (1989). *The Unheeded Cry: Animal Consciousness, Animal Pain and Scientific Change*. Oxford University Press.

ROLSTON, H. (1983). Values gone wild. *Inquiry*, 181–207.

ROUTLEY, V. (1975). Critical notice of John Passmore's 'Man's Responsibility for Nature'. *Austral. J. Phil.*, **53**, 171–85.

ROWAN, A. (1986). *Animal awareness. Are animals anxious?* delivered at Delta Soc. Congress Living together; people, animals and the environment.

SALIH, M. S. (1978). Economics and problems of rangeland productivity in Sudan. In D. N. Hyder (ed.), *Proceedings of the*

First International Rangeland Congress. Society of Range Management, Denver.

SAPONSIS, S. F. (1987). *Morals, Reasons and Animals*. Temple University Press, Philadelphia.

SCHENKEL, R. (1947). *Ausdrucks—studien an Wolfen Behav.*, **1**, 81–129.

SCHUMACHER, E. F. (1974). *Small is Beautiful*. Abacus, London.

SELYE, H. (1950). *The Physiology and Pathology of Exposure to Stress*. Acta, Montreal.

SERPELL, J. (1986). *In the Company of Animals*. Blackwell, Oxford.

SHANE, D. (1987). Assault on Eden; destruction of Latin America's rain forests. In M. W. Fox and L. D. Mickley (eds.), *Advances in Animal Welfare Science 1986/87*, pp. 149–64 (op. cit.).

SHARPE, R. (1987). The Cruel Deception. In M. W. Fox and L. D. Mickley (eds.), *Advances in Animal Welfare Science 1986/87* (op. cit.).

SHOARD, M. (1980). *The Theft of the Countryside*. Temple Smith, London.

SINGER, P. (1976). *Animal Liberation*. Jonathan Cape, London.

SMITH, R. F. and REYNOLDS, H. T. (1966). Principles, definition and scope of integrated pest control. *Proc. FAO Symp. Integrated Pest Control*, **1**, 11–17.

SPEDDING, C. R. W., WALSINGHAM, J. M. and HOXEY, A. M. (1981). *Biological Efficiency in Agriculture*. Academic Press, London.

STAMP-DAWKINS, M. (1980). *Animal Suffering. The Science of Animal Welfare*. Chapman & Hall, London.

STAMP-DAWKINS, M. (1983). Battery hens have their price; consumer demand theory and the measurement of 'ethological needs'. *Anim. Behav.*, **31**, 1195–1205.

STEINHART, O. and STEINHART, J. S. (1974). Energy use in the US food system. *Science*, **194**, 307–16.

STERN, V. M. R., MUELLER, V., SEVADARIAN, V. and WAY, M. (1969). Lygus bug control in cotton through alfalfa interplanting. *Calif. Agric.*, **23**, 8–10.

STOLBA, A. (1982). A family system for pig housing. In *UFAW Alternatives to Intensive Husbandry Systems*. UFAW Symp. Wye College, Ashford, Kent, pp. 52–67.

SWANN REPORT (1969). *The Use of Antibiotics in Agriculture*. HMSO, London.

TANSLEY, A. G. (1945). *Our Heritage of Wild Nature—A Plea for Organised Nature Conservation*. Oxford University Press.

TEWFIK, M. (1977). A pilot project for the Khors, Eastern Sudan. *Ekistics*, **43**, 258.

THOMAS, K. (1984). *Man and the Natural World. Changing attitudes in England 1500–1800*. Penguin, Harmondsworth.

THORPE, W. H. (1965). The assessment of pain and distress in animals. In *Report of the Technical Committee to Enquire into the Welfare of Animals kept under Intensive Livestock Systems*. Chairman F. W. R. Brambell. 2836. HMSO, London.

TOLMAN, E. C. and HONZIK. C. H. (1930). *Introduction and removal of reward, and maze performance in rats*. Univ. of California. Publications in Psychology, **4**, 257–75.

TOTHILL, J. D. (1948). *Agriculture in the Sudan*. Oxford University Press.

TURNBULL, C. (1980). *The Forest People*. Paladin, London.

TURNBULL, C. (1984). *The Mountain People*. Paladin, London.

UFAW (1981). *Alternatives to Intensive Animal Husbandry Systems*. Universities Federation for Animal Welfare; Symposium held at Wye College, Ashford, Kent.

UFAW (1983). *Symposium on Animal Awareness*. Oxford.

ULBRICHT, T. (1980). In R. Boeringa (ed.), Alternative Methods of Agriculture. *Agriculture Environment*, **5**, v–vi.

ULRICH, R. (1966). Pain as a cause of aggression. *Amer. Zool.*, **6**, 643–62.

UNIDO (1976). *Industrialisation of the Least Developed Countries*. Report of the Intergovernmental Expert Group Meeting, Vienna, 15–24 November 1976, New York.

UNIVERSITY OF READING (1982). *Farm Business Data*. Dept. of Agricultural Economics and Management, University of Reading.

US DEPT. OF AGRICULTURE (1980). *Report and Recommendations on Organic Farming*. 1980-0-310-944/96. US Government Printing Office, Washington, DC.

VALE, R. (1977). *Smallholding and Food Production*. Alternative Technology Group, The Open University, Milton Keynes.

VINE, A. and BATEMAN, D. (1982). Some economic aspects of organic farming in England and Wales. *Biological Agriculture and Horticulture*, **1**, 65–72.

VON VEXHULL, J. (1934). A walk through the worlds of animals and men. In C. H. Schiller (ed.), *Instinctive Behaviour.* New York University Press.

WALKER, S. (1983). *Animal Thought.* Routledge & Kegan Paul, London.

WAY, M. J. (1977). Integrated control—practical realistics. *Outlook Agriculture*, **9**, 127–35.

WHITE, R. F. (1981). *Introduction to the Principles and Practice of Soil Science.* Blackwell, Oxford.

WILSON, E. O. (1975). *Sociobiology.* Belknap Press, Cambridge, Massachusetts.

WILSON, R. J. A. and DORAN, T. (1980). The Kenana Sugar Project in Sudan. *Agribusiness Worldwide*, 1.

WILSON, R. T. (1976). Studies on the livestock of southern Darfur, Sudan. II. Production traits in cattle. *Tropical Animal Health and Production*, **8**, 1.

WILSON, R. T. (1978). Studies on the livestock of southern Darfur, Sudan. VI. Notes on equines. *Tropical Animal Health and Production*, **10**, 3.

WOODS, A. (1974). *Pest Control. A Survey.* Cambridge University Press.

WORTHINGTON, E. B. (1946). *Middle East Science.* HMSO, London.

WORTHINGTON, S. and WORTHINGTON, E. B. (1933). *Inland Waters of Africa.* Macmillan, London.

ZAHLAN, A. B. (ed.) (1984). *Agricultural Bibliography of the Sudan 1974–83.* Ithaca Press, London.

INDEX